Prentice Hall
Reference to Mathematics

A Guide for Everyday Math

Cheryl Cleaves
Southwest Tennessee Community College

Margie Hobbs
The University of Mississippi

Upper Saddle River, New Jersey
Columbus, Ohio

Library of Congress Cataloging-in-Publication Data

Cleaves, Cheryl S., (date)
  Prentice Hall's reference to mathematics : a guide for everyday math / Cheryl Cleaves,
  Marge Hobbs.
    p. cm.
  Includes index.
    1. Mathematics—Handbooks, manuals, etc. I. Title: Reference to mathematics. II.
  Hobbs, Margie, J., (date)- III. Title.

  QA40 .C54 2003
  510—dc21
                                                                        2002074903

Editor in Chief: Stephen Helba
Executive Editor: Frank I. Mortimer, Jr.
Production Editor: Louise N. Sette
Production Supervision: Clarinda Publication Services
Design Coordinator: Diane Ernsberger
Cover Designer: Mark Shumaker
Production Manager: Brian Fox
Marketing Manager: Tim Peyton

This book was set in Sabon and Frutiger by The Clarinda Company. It was printed
and bound by R. R. Donnelley & Sons Company. The cover was printed by Phoenix
Color Corp.

Pearson Education Ltd.
Pearson Education Australia Pty. Limited
Pearson Education Singapore Pte. Ltd.
Pearson Education North Asia Ltd.
Pearson Education Canada, Ltd.
Pearson Educatión de Mexico, S.A. de C.V.
Pearson Education—Japan
Pearson Education Malaysia Pte. Ltd.
Pearson Education, *Upper Saddle River, New Jersey*

10 9 8 7 6 5 4 3 2 1
ISBN: 0-13-061800-4

How to Use
Prentice Hall's Reference to Mathematics: A Guide for Everyday Math

Every home or workplace should have a mathematics reference just as it has a dictionary or language handbook. Too often we need to use a mathematical procedure that we haven't used in a long time only to realize that we just remember bits and pieces of the procedure. *Prentice Hall's Reference to Mathematics* is designed to help you fill in those gaps. We have developed this resource in response to the many requests we have received from our colleagues, friends, and family who are students, parents, working adults, and teachers. They are often frustrated with the limited resources they have available when they try to learn or review a mathematical process.

Finding What You Need

For a reference to be useful, you must be able to find what you need. We provide several different strategies for finding topics in this book. Different strategies are for different purposes, and today you may use one strategy and tomorrow another.

- The **Table of Contents** shows how the topics are organized and for many topics it gives you a logical progression from one concept to another. Topics are grouped in major categories: *Computations, Equations and Formulas, Measurement and Geometry,* and *Statistics and Probability.* Each major category is divided into chapters, sections, and specific topics. Each chapter opens with a list of sections included in the chapter.

  You may use the Table of Contents when you want to review several related topics and to become familiar with the organization of the text.

- The *index* portion of the **Glossary/Index** helps you locate topics when you remember the terminology associated with the concept. Since the glossary and index are merged, the *glossary* portion lets you quickly confirm the definition of a term. Together these two tools help you to locate prerequisite skills needed for a topic, related topics, and follow-up topics.

Related Topics ─┐
Application ─┤

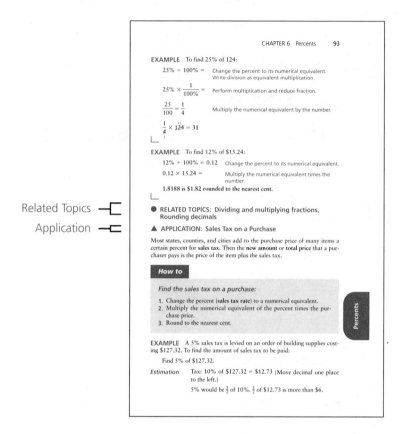

EXAMPLE To find 25% of 124:

25% ÷ 100% =   Change the percent to its numerical equivalent. Write division as equivalent multiplication.

$25\% \times \dfrac{1}{100\%} =$ Perform multiplication and reduce fraction.

$\dfrac{25}{100} = \dfrac{1}{4}$ Multiply the numerical equivalent by the number.

$\dfrac{1}{4} \times 124 = 31$

EXAMPLE To find 12% of $15.24:

12% ÷ 100% = 0.12   Change the percent to its numerical equivalent.

0.12 × 15.24 =   Multiply the numerical equivalent times the number.

1.8188 is $1.82 rounded to the nearest cent.

● **RELATED TOPICS:** Dividing and multiplying fractions, Rounding decimals

▲ **APPLICATION:** Sales Tax on a Purchase

Most states, counties, and cities add to the purchase price of many items a certain percent for **sales tax**. Then the **new amount** or **total price** that a purchaser pays is the price of the item plus the sales tax.

How to

Find the sales tax on a purchase:

1. Change the percent (**sales tax rate**) to a numerical equivalent.
2. Multiply the numerical equivalent of the percent times the purchase price.
3. Round to the nearest cent.

EXAMPLE A 5% sales tax is levied on an order of building supplies costing $127.32. To find the amount of sales tax to be paid:

Find 5% of $127.32.

Estimation   Tax: 10% of $127.32 = $12.73 (Move decimal one place to the left.)

5% would be $\frac{1}{2}$ of 10%. $\frac{1}{2}$ of $12.73 is more than $6.

Percents

● **Related Topics** are listed whenever appropriate. These topics refer to prerequisite skills and follow-up topics as well as related topics.

▲ **Applications** include many business, career, workplace, and consumer topics that use various mathematical skills. These topics can help you answer questions such as "How much money do I need to set aside each month to prepare for my child's education or my retirement?" or "How much will my payment be if I finance $20,000 for 15 years at 7% interest?" An alphabetized list of applications is included on the inside covers.

Understanding What You Read

Locating a topic is only the first step in meeting your needs. You need to understand the explanation given for the procedure. Both of us have taught mathematics for over 35 years at the college, high school, and middle school or junior high levels. In addition, we have written numerous mathematics textbooks. We try especially hard to make our texts readable. Persons who need to learn or review mathematics are often very good readers. Mathematically precise terminology is used so you will have the key words for referencing other resources, and casual and informal language is used to enhance your understanding. A variety of features is presented to accommodate various learning styles.

- **Definitions, rules, properties, and procedures** are included using mathematical terminology and step-by-step instructions. Key words are in **boldface** type and the definitions are included in the narrative. Many of the definitions are repeated in the glossary. Rules, properties, and procedures are in **How to** boxes. Procedures and mathematical properties are also written symbolically when appropriate. This will help you make the transition to other resources that use only the symbolic representation.

- **Examples with explanatory notes** follow each **How to** box. The explanatory notes reinforce the steps used to solve the problem. Color shading is used within the solution of many examples to help you follow the path of key values.

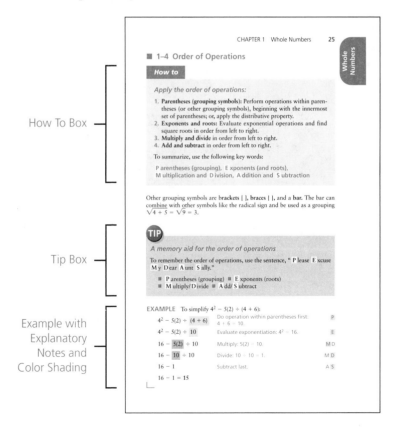

How To Box

Tip Box

Example with Explanatory Notes and Color Shading

CHAPTER 1   Whole Numbers   25

Whole Numbers

■ 1–4 Order of Operations

How to

Apply the order of operations:

1. **Parentheses (grouping symbols):** Perform operations within parentheses (or other grouping symbols), beginning with the innermost set of parentheses; or, apply the distributive property.
2. **Exponents and roots:** Evaluate exponential operations and find square roots in order from left to right.
3. **Multiply and divide** in order from left to right.
4. **Add and subtract** in order from left to right.

To summarize, use the following key words:

P arentheses (grouping), E xponents (and roots),
M ultiplication and D ivision, A ddition and S ubtraction

Other grouping symbols are **brackets [ ], braces { },** and a **bar.** The bar can combine with other symbols like the radical sign and be used as a grouping $\sqrt{4 + 5} = \sqrt{9} = 3.$

TIP

A memory aid for the order of operations

To remember the order of operations, use the sentence, " P lease E xcuse M y D ear A unt S ally."

■ P arentheses (grouping) ■ E xponents (roots)
■ M ultiply/ D ivide ■ A dd/ S ubtract

EXAMPLE   To simplify $4^2 - 5(2) \div (4 + 6)$:

$4^2 - 5(2) \div (4 + 6)$	Do operation within parentheses first: $4 + 6 = 10.$	P
$4^2 - 5(2) \div 10$	Evaluate exponentiation: $4^2 = 16.$	E
$16 - 5(2) \div 10$	Multiply: $5(2) = 10.$	M D
$16 - 10 \div 10$	Divide: $10 \div 10 = 1.$	M D
$16 - 1$	Subtract last.	A S
$16 - 1 = 15$		

- The **Tip** boxes answer many of the questions that you may have. Each tip box has a title so that you can quickly determine the content of the tip. These tips give shortcut procedures or cautions that you may want to know.
- The **Six-Step Problem Solving Plan** is a systematic approach to solving applied problems. The structured approach helps you develop intuitive problem-solving skills. This plan is introduced on p. ix and examples are given throughout.

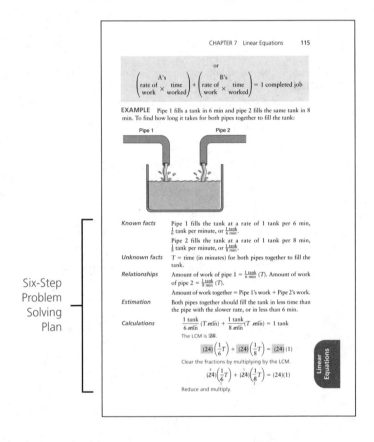

Acknowledgments

We wish to thank the Prentice Hall team who helped make this book a reality. In particular we thank Steve Helba, Editor-in-Chief, who believed in our ideas and supported our work. Frank Mortimer, Executive Editor, provided the vision and leadership throughout the development and production of the project. Tim Peyton was most helpful in providing research on the need for a mathematical reference. We also thank Louise Sette, production editor. Emily Autumn of Clarinda Publication Services was especially helpful.

A comprehensive glossary/index is an essential component of any reference book. We are indebted to Shirley Riddle for creating the extensive, integrated glossary and index for this book.

We dedicate this book to our friend and colleague Ada Shotwell. She has been a source of encouragement and inspiration in our professional lives.

We thank our families, especially Charles Cleaves and Allen and Holly Hobbs, for their love and support.

We have devoted our entire careers helping others develop confidence in their mathematical abilities. It is important that you feel empowered to accomplish whatever you want without fear that mathematics will hold you back. If this text helps you gain this empowerment, even in a small way, all the work will have been worth it. We value your input because it helps us to continue to improve our products. We welcome your comments and suggestions; you can email us at ccleaves@bellsouth.net or mhobbs@watervalley.net.

Cheryl Cleaves
Margie Hobbs

Problem-Solving Strategies

Problem solving is an important skill in the workplace and in everyday life. In developing good problem-solving skills it is helpful to use a systematic problem-solving plan. A plan gives you a framework for approaching problems.

Develop Problem-Solving Skills with a Positive Attitude

It is important to keep a positive attitude when solving problems.

Work with a Study Partner or Group

Working with a partner or group develops confidence. As you work with others, each of you verbalizes your understanding of the problem, and any misconceptions you might have can be discussed and corrected.

You and your partners can examine several approaches to the problem and discuss their pros and cons. This discussion strengthens your investigative skills and gives you more options than you would probably come up with alone. And, you may discover that there may be several *correct* approaches to solving the problem.

Be Persistent

Even the best problem solvers do not always use a correct approach on the first try. The mark of a good problem solver is persistence. Learn from failed attempts. Learn what doesn't work and why. This makes the correct approach easier to understand when you find it.

Don't Make the "Ruts" Too Deep

After pursuing several dead ends, don't continue to the point of frustration. In other words, don't spin your wheels! The harder you work, the deeper the ruts get. Back away from the problem, and when you come back to it, you may think of a new approach you haven't tried before.

Or, seek help from others (additional classmates, tutors, your instructor). Just because a problem doesn't "click" right away, doesn't mean that you can't solve problems.

You will encounter many different plans for solving problems. They are all trying to help you to organize the problem details so that you can find an appropriate procedure for solving the problem. Our plan, like others, is very structured. As you develop confidence and skill in solving problems, you will probably use less structured and more intuitive approaches.

Six-Step Problem-Solving Plan

1. **Unknown Facts.** What facts are missing from the problem? What are you trying to find?
2. **Known Facts.** What relevant facts are known or given? What facts must you bring to the problem from your own background?
3. **Relationships.** How are the known facts and the unknown facts related? What formulas or definitions can you use to establish a model?
4. **Estimation.** What are some of the characteristics of a reasonable solution? For instance, should the answer be more than a certain amount or less than a certain amount?
5. **Calculations.** Perform the operations identified in the relationships.
6. **Interpretation.** What do the results of the calculations represent within the context of the problem? Is the answer reasonable? Have all unknown facts been found? Do the unknown facts check in the context of the problem?

EXAMPLE   The 7th Inning buys baseball cards from eight different vendors. In November, the company purchased 8,832 boxes of cards. If an equal number of boxes was purchased from each vendor, how many boxes of cards were supplied by each vendor?

Unknown fact   Number of boxes of cards supplied by each vendor

Known facts   8,832   total number of boxes purchased
8   total number of vendors
An equal number of boxes were purchased from each vendor.

Relationships   Boxes purchased by each vendor =
Total boxes purchased ÷ Number of vendors

Estimation   If 8,000 boxes were purchased in equal amounts from 8 vendors, then 1,000 were purchased from each vendor. Since more than 8,000 boxes were purchased, then more than 1,000 were purchased from each vendor.

Calculations $\dfrac{1{,}104}{8\overline{)8{,}832}}$

Interpretation   Each vendor supplied 1,104 boxes of cards.

└

In identifying the relationships of a problem or in developing your plan for solving a problem, it is often helpful if you look for key words or phrases that give you clues about the mathematical operations involved in the relationships. Key words give clues as to whether one quantity is added to, subtracted from, or multiplied or divided by another quantity. For example, if a problem tells you that Carol's salary in 2003 exceeds her 2002 salary by $2,500, you know that you should add $2,500 to her 2002 salary to find her 2003 salary.

Key Words and What They Generally Imply in Word Problems

Addition	Subtraction	Multiplication	Division	Equality
The sum of	Less than	Times	Divide(s)	Equals
Plus/total	Decreased by	Multiplied by	Divided by	Is/was/are
Increased by	Subtracted	Of	Divided into	Is equal to
More/more than	from	The product of	Half of (divided by 2)	The result is
Added to	Difference between	Twice	Third of	What is left? What remains?
Exceeds	Diminished by	(2 times)	(divided by 3)	The same as
Expands	Take away	Double	Per	Gives/giving
Greater than	Reduced by	(2 times)	How big is each part?	Makes
Gain/profit	Less/minus	Triple	How many parts can be made from . . .?	Leaves
Longer	Loss	(3 times)		
Older	Lower	Half of ($\frac{1}{2}$ times)		
Heavier	Shrinks	Third of ($\frac{1}{3}$ times)		
Wider	Smaller than			
Taller	Younger			
	Slower			

In many applied problems, you must use more than one relationship to find the unknown facts. When this is the case, you usually need to read the problem several times to find all the relationships and plan your solution strategy.

EXAMPLE   Carlee Anne McAnally needs to ship 78 crystal vases. With standard packing to avoid damage, 5 vases fit in each available box. How many boxes are required to pack the vases?

Unknown fact   Number of boxes required to ship the vases

Known facts   Total vases to be shipped: 78
Number of vases per box: 5

Relationships	Total boxes needed = Total number of vases ÷ number per box Total boxes needed = 78 ÷ 5 If there is a remainder, an extra box, that will be partially full, will be needed.
Estimation	70 ÷ 5 = 14 Round down the dividend. 80 ÷ 5 = 16 Round up the dividend. Since 78 is between 70 and 80, the number of boxes needed is between 14 and 16.
Calculation	78 ÷ 5 = 15 R 3
Interpretation	**16 boxes are needed;** 15 boxes will contain 5 vases each, and 1 box will contain 3 vases. The box with 3 vases will need extra packing.

TIP

Using guess and check to solve problems

A strategy that is often effective for solving problems involves guessing. Make a guess that you think might be reasonable and check to see if the answer is correct. If your guess is not correct, decide if it is too high or too low. Make another guess based on what you learned from your first guess. Continue until you find the correct answer.

Let's try guessing in the previous example. We found that we could pack 70 vases in 14 boxes and 80 vases in 16 boxes. Since we need to pack 78 vases, how many vases can we pack with 15 boxes? 15 × 5 = 75. Still not enough. Therefore, we will need 16 boxes but the last box will not be full.

You can probably think of other ways to solve this problem. This is good. Any plan that leads to a correct solution is acceptable. Some plans will be more efficient than others, but you develop your problem-solving skills by pursuing a variety of strategies.

Table of Contents

CHAPTER **2**

Fractions   27

CHAPTER *3*

Decimals   59

CHAPTER *4*

Integers and Signed Numbers   71

CHAPTER **8**

Graphing Linear Equations   137

CHAPTER **9**

Slope and Distance   150

CHAPTER *13*

Products and Factors   206

CHAPTER **17**
Exponential and Logarithmic Equations   266

PART **3**

Measurement and Geometry   285

CHAPTER **18**

Direct Measurement   286

PART **4**

Statistics and Probability   367

CHAPTER **21**

Interpreting and Analyzing
Data   368

Appendix

Glossary/Index

Computations

Whole Numbers

▓ 1–1 Whole Numbers and the Place-Value System

Our system of numbers, the **decimal-number system** uses 10 symbols called **digits:** 0, 1, 2, 3, 4, 5, 6, 7, 8, 9. A **whole number** is made up of one or more digits.

Each place a digit occupies in a number has a value called a **place value.** Each place value *increases* as we move from *right* to *left* and each increase is *10 times* the value of the place to the right (Fig. 1–1).

Billions			Millions			Thousands			Units		
Hundred billions (100,000,000,000's)	Ten billions (10,000,000,000's)	Billions (1,000,000,000's)	Hundred millions (100,000,000's)	Ten millions (10,000,000's)	Millions (1,000,000's)	Hundred thousands (100,000's)	Ten thousands (10,000's)	Thousands (1,000's)	Hundreds (100's)	Tens (10's)	Ones (1's)

Figure 1–1   Whole-number place values and periods.

The place values are arranged in **periods,** or groups of three. The first period is called **units,** the second period is called **thousands,** and so on. Each period has a hundreds place, a tens place, and a ones place.

In four-digit numbers, the comma separating the units group from the thousands group is optional. Thus, 4,575 and 4575 are both correct.

How to

Identify the place value of digits:

1. Mentally position the number on the place-value chart so the last digit on the right aligns under the ones place.
2. Identify the place value of each digit according to its position on the chart.

EXAMPLE To identify the place value of the digit 7 in the number 2,472,694,500:

Billions			Millions			Thousands			Units		
Hundred billions	Ten billions	Billions	Hundred millions	Ten millions	Millions	Hundred thousands	Ten thousands	Thousands	Hundreds	Tens	Ones
		2	4	7	2	6	9	4	5	0	0

1. Position number on chart.

Figure 1–2 Whole-number place-value chart.

ten-millions place

2,	472,	694,	500
Billions	Millions	Thousands	Units

2. Identify place value of 7.

7 is in the ten-millions place.

RELATED TOPICS: Decimals, Metric system of measurement, Power of 10, Prefixes, Scientific notation

How to

Read or write whole numbers:

1. Mentally position the number on the place-value chart so the last digit on the right aligns under the ones place.
2. Examine the number from right to left, separating each period with commas.
3. Identify the leftmost period.
4. From the left, read the numbers in each period and the period name. (The units period is not usually read. A period containing all zeros is not usually read.)

EXAMPLE The number 2472694500 would be read in words as:

2,472,694,500 Separate with commas.

two *billion,* four hundred seventy-two *million,* six hundred ninety-four *thousand,* five hundred.

Special conventions with reading and writing whole numbers

- A period name is inserted before each comma.
- The word *and* should not be used when reading whole numbers.
- The numbers from 21 to 99 (except 30, 40, 50, and so on) use a hyphen when they are written (twenty-six, forty-three, and so on).

● **RELATED TOPIC: Decimals**

Numbers written with digits in the appropriate place-value positions are in **standard notation**.

How to

Write a whole number in standard notation:

1. Write the period names in order from left to right, starting with the first period in the number.
2. Place the digits in each period. Leave blanks for zeros if necessary.
3. Insert zeros as needed for each period. Each period except the leftmost period must have three digits.
4. Separate the periods with commas.

EXAMPLE   In standard notation, eight million, two hundred four thousand, twelve is written as:

Millions	Thousands	Units	Write period names in order.
8	204	12	Place digits in each period.
8	204	0 12	Insert zeros as needed.
8,	204,	012	Separate periods with commas.

8,204,012.

Expanded notation shows the sum of each digit times its place value.

How to

Write a whole number in expanded notation:

Write 516 in expanded notation.

1. Write each digit times its place value.

5 × 100
1 × 10
6 × 1

2. Write the expanded digits as indicated addition.

(5 × 100) + (1 × 10) + (6 × 1)

EXAMPLE   In expanded notation 2,305 is written as:

(2 × 1,000) + (3 × 100) + (0 × 10) + (5 × 1)

or (2 × 1,000) + (3 × 100) + (5 × 1)   A place value of zero does not have to be written.

● **RELATED TOPICS:** Grouping, Order of operations, Power of 10

Rounding a number means finding the closest **approximate number** to a given number. For example, if 37 is rounded to the nearest ten, is 37 closer to 30 or 40? Locate 37 on the number line.

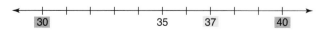

37 is closer to 40 than 30. Thus, 40 is a better approximation to the nearest ten for 37. Another way to say this is that 37 rounded to the nearest ten is 40.

How to

Round a whole number to a given place value:

1. Locate the digit in the rounding place. Then examine the digit to the immediate right.
2. If the digit to the right of the rounding place is 0, 1, 2, 3, or 4, do not change the digit in the rounding place. If the digit to the right of the rounding place is 5, 6, 7, 8, or 9, add 1 to the digit in the rounding place.
3. Replace all digits to the *right* of the digit in the rounding place with zeros.

EXAMPLE   326 rounded to the hundreds place is:

3 2 6   3 is in the hundreds place. The next digit to the right is 2.

3 2 6   Leave 3 unchanged because 2 is less than 5.

3

300   Replace 2 and 6 with zeros.

EXAMPLE   46,897 rounded to the tens place is:

46,8 9 7   9 is in the tens place and the next digit to the right is 7.

46,8 9 7   Add 1 to 9 to give 10. Place zero in the tens place and add 1 to the 8 in the hundreds place. (89 + 1 = 90)

46,900   Replace 7 with zero.

Nine plus one still equals ten

When the digit in the rounding place is 9 and must be rounded up, it becomes 10. The 0 replaces the 9 and 1 is carried to the next place to the left.

● **RELATED TOPICS: Estimation, Rounding decimals, Significant digits**

Whole numbers can be arranged on a **number line** to show a visual representation of the relationship of numbers by size. The most common arrange-

ment is to begin with zero and place numbers on the line from left to right as they get larger.

All numbers have a place on the number line and the numbers continue indefinitely without end. A term that is often used to describe this concept is **infinity** and the symbol is ∞.

Whole numbers can be compared by size by determining which of the two numbers is larger or smaller. If two numbers are positioned on a number line, the smaller number is positioned to the left of the larger number. The order relationship can be written in a mathematical statement called an **inequality**. An inequality shows that two numbers are not equal; that is, one is larger than the other. Symbols for showing inequalities are the **less than** symbol < and the **greater than** symbol >.

$5 < 7$ Five is less than seven.

$7 > 5$ Seven is greater than five.

How to

Compare whole numbers:

1. Mentally position the numbers on a number line.
2. Select the number that is farther to the left to be the smaller number.
3. Write an inequality using the *less than* symbol.

   smaller number < larger number

   or

   Write an inequality using the *greater than* symbol.

   larger number > smaller number

Which way does the inequality symbol point?

In using an inequality symbol to relate two numbers, the point of the *less than* symbol is directed toward the *left* like the arrowhead on the *left* of the number line. The point of the *greater than* symbol is directed toward the *right* like the arrowhead on the *right* of the number line.

$3 < 5$ $6 < 14$ $4 > 2$ $9 > 7$

EXAMPLE   To write an inequality comparing the numbers 12 and 19:

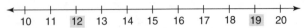

Mentally position the numbers on the number line.

12 is the smaller number   12 is to the *left* of 19.

12 < 19 or **19 > 12**   Use appropriate inequality symbol.

● **RELATED TOPICS: Comparing decimals, Comparing fractions, Comparing integers, Linear inequalities**

Numbers are used to show *how many* and to show *order*. **Cardinal numbers** show *how many* and **ordinal numbers** show *order* or position (such as first, second, third, fourth, etc.). For example, in the statement "three students are doing a presentation," three is a cardinal number (showing how many). In the statement "Margaret is the third tallest student in the class," third is an ordinal number (showing order).

   Front end rounding and **rounding to one nonzero digit** both imply the same process with respect to whole numbers.

How to

Round a whole number to one nonzero digit:

1. Locate the leftmost digit (front end digit). This digit is in the rounding place. Then examine the digit to the immediate right.
2. If the digit to the right of the rounding place is 0, 1, 2, 3, or 4, do *not* change the digit in the rounding place. If the digit to the right of the rounding place is 5, 6, 7, 8, or 9, add 1 to the digit in the rounding place.
3. Replace *all* digits to the *right* of the rounding place with zeros.

EXAMPLE   To round 3892 using front end rounding or rounding to one nonzero digit:

3 892   Identify the rounding place which is the leftmost place in the number. 3 is the digit in the rounding place.

3 8 92   The digit to the right of the rounding place is 8. So add 1 to 3 (sometimes called **rounding up**).

4 000   Replace digits on the right of the rounding place with zeros.

   The **significant digits** of a whole number are all nonzero digits and all digits that are zero if the zero is between two nonzero digits. The significant digits of 820,500 are 8, 2, 0, and 5. The two zeros on the end are *not* considered significant digits. Thus, 820,500 has 4 significant digits.

How to

Round a whole number to a specified number of significant digits:

1. Determine the number of significant digits that are desired in the rounded number.
2. Locate the digit, counting from *left* to *right* that is the same number of digits from the left (front end of the number) as the desired number of significant digits. This digit is in the rounding place.
3. If the digit to the right of the rounding place is 0, 1, 2, 3, or 4, do *not* change the digit in the rounding place. If the digit to the right of the rounding place is 5, 6, 7, 8, or 9, add 1 to the digit in the rounding place.
4. Replace *all* digits to the *right* of the rounding place with zeros.

EXAMPLE   To round the number 820,500 to three significant digits:

82 0 ,500	The third digit from the left is identified as being in the rounding place.
82 0 , 5 00	The digit to the right of 0 is 5 so 1 is added to 0.
821,000	The digits to the right of the rounding place are replaced with zeros and the digits to the left of the rounding place stay the same.

● **RELATED TOPIC: Rounding Decimals**

■ 1–2 Basic Operations with Whole Numbers

Numbers being added are called **addends,** and the answer is called the **sum** or **total.** Addition is **commutative.** That means numbers can be added in any **order.**

$$7 + 6 = 13 \quad \text{or} \quad 6 + 7 = 13$$

Addition is **associative.** That means that numbers being added may be **grouped** in any manner.

$$(7 + 4) + 6 = 11 + 6 = 17 \quad \text{or} \quad 7 + (4 + 6) = 7 + 10 = 17$$

Addition is a **binary operation;** that is, the rules of addition apply to adding *two* numbers at a time.

Using symbols to write rules and definitions

Many rules and definitions can be written symbolically. Symbolic representation provides a visual recognition of the rule or definition.

Commutative Property of Addition: Two numbers may be added in any order and the sum remains the same.

$a + b = b + a$ where a and b are numbers.

Associative Property of Addition: Three numbers may be added using different groupings and the sum remains the same.

$a + (b + c) = (a + b) + c$ where a, b, and c are numbers.

The associative property of addition also allows other possible groupings and extends to more than three numbers.

$$7 + 4 + 6 = 13 + 4 = 17$$
$$3 + 5 + 7 + 9 = 8 + 16 = 24$$

● **RELATED TOPICS: Associative property of multiplication, Commutative property of multiplication, Subtraction**

Adding zero to any number results in the same number. This property is called the **zero property of addition** and zero is called the **additive identity.**

$n + 0 = n$ or $0 + n = n$
$5 + 0 = 5$ or $0 + 5 = 5$

How to

Add numbers of two or more digits:

1. Arrange the numbers in columns so that the ones place values are in the same column.
2. Add the ones column, then the tens column, then the hundreds column, and so on, until all the columns have been added. **Carry** whenever the sum of a column is more than one digit. This is also called **regrouping.**

EXAMPLE   The sum of 250, 75, 12, and 8 is:

$$
\begin{array}{r}
\overset{1\,1}{2}5 0 \\
7 5 \\
1 2 \\
+\quad 8 \\
\hline
3 4 5
\end{array}
$$

The sum of the digits in the ones column is 15. Record the 5 in the ones column and *carry* the 1 to the tens column.
The sum of the digits in the tens column is 14. Record the 4 in the tens column and carry the 1 to the hundreds column.

The sum is 345.

● **RELATED TOPIC: Adding integers**

How to

Estimate the sum of an addition problem:

1. Round each addend to a specific place value or to a number with one nonzero digit.
2. Add the rounded addends.

● **RELATED TOPICS: One nonzero digit, Rounding whole numbers**

How to

Check an addition problem:

1. Add the numbers a second time and compare with the first sum.
2. Use a different order or grouping if convenient.

EXAMPLE   To find the total of $16,466, $23,963, and $5,855, first estimate by rounding to thousands. Then, find the exact amount and check.

Thousands Place	Estimate	Exact	Check
$1 6 ,466	$16,000	$16,466	$16,466
2 3 ,963	24,000	23,963	23,963
5 ,855	6,000	5,855	5,855
	$46,000	$46,284	$46,284

The estimate and exact sum are close. The exact answer is reasonable.

Accuracy of estimates

The accuracy of estimates varies depending on the place value to which the addends are rounded. The context of a problem will help in determining a reasonable rounding place.

Subtraction is the **inverse operation** of addition. To solve the subtraction problem $9 - 5 = ?$ we ask, "What number must be added to 5 to give 9?" The answer is 4 because 4 added to 5 gives a total of 9. The result of subtraction is called the **difference** or **remainder**. The initial quantity is the **minuend**. The amount being subtracted is the **subtrahend**.

Subtraction is not commutative. $8 - 3 = 5$, but $3 - 8$ does not equal 5; that is $3 - 8 \neq 5$. The symbol $\neq$ is read "is not equal to."

Subtraction is not associative.

$$9 - (5 - 1) = 9 - 4 = 5 \quad \text{but} \quad (9 - 5) - 1 = 4 - 1 = 3$$

$$5 \neq 3$$

What do you do if there are no groupings when subtracting more than two numbers?

If no grouping symbols are included, perform subtractions from left to right since subtraction is *not* associative.

Subtract $8 - 3 - 1$.

$$8 - 3 - 1 = 5 - 1 = 4$$

● **RELATED TOPICS: Associative property of addition, Commutative property of addition**

Subtracting zero from a number results in the same number:

$$n - 0 = n, \qquad 7 - 0 = 7$$

How to

Subtract numbers of two or more digits:

1. Arrange the numbers in columns, with the minuend at the top and the subtrahend at the bottom.
2. Make sure the ones digits are in a vertical line on the right.
3. Subtract the digits in the ones column first, then the tens column, the hundreds column, and so on.
4. To subtract a larger digit from a smaller digit in a column, **borrow** 1 from the digit in the next column to the left. This is the same as borrowing *one* group of 10; thus, add 10 to the digit in the given column. Then, continue subtracting.

The concept of **borrowing** is also referred to as **regrouping**.

EXAMPLE   To subtract 9,327 − 3,514:

$$
\begin{array}{r}
\overset{8\ \ 13}{9,\ 327} \\
-\ 3,\ 514 \\
\hline
5,\ 813
\end{array}
$$

Arrange in columns.
In the hundreds place 5 is more than 3. Borrow 1 group of 10 from 9, 9 − 1 = 8, 10 + 3 = 13.

● **RELATED TOPIC: Subtracting integers**

How to

Estimate the difference:

1. Round each number to the indicated place value or to a number with one nonzero digit.
2. Subtract the rounded numbers.

How to

Check a subtraction problem:

1. Add the subtrahend and difference.
2. Compare the result of Step 1 with the minuend. If the two numbers are equal, the subtraction is correct.

EXAMPLE   To find the difference of 427 − 125, first estimate by rounding to hundreds. Then, the exact difference is found. The answer checks.

	Estimate	Exact	Check
4 27	400	427	125
− 1 25	− 100	− 125	+ 302
	300	302	427

∟

  Multiplication is repeated addition. If we have three $10 bills, we have
$10 + $10 + $10 = $30 or 3 × $10 = $30.
  The numbers in multiplication are called **factors.** Sometimes the first
number is called the **multiplicand,** and the second number is called the
multiplier.
  The answer or result of multiplication is called the **product.**

2	×	3	=	6
factor		factor		product
or		or		
multiplicand		multiplier		

TIP

Various notations for multiplication

Besides the familiar × or "times" sign, the raised dot (·), the asterisk (∗),
and parentheses ( ) are also used to show multiplication.

  2 · 3 = 6,   2 ∗ 3 = 6,   2(3) = 6,   (2)(3) = 6

  The **commutative property of multiplication** permits two numbers to be
multiplied in any **order.** In symbols, $a \times b = b \times a$.

  4 × 5 = 20,   5 × 4 = 20

  The **associative property of multiplication** permits more than two numbers to be **grouped** in any way. In symbols, $a \times (b \times c) = (a \times b) \times c$.

2 × (3 × 5)	or	(2 × 3) × 5
2 × 15		6 × 5
30		30

● **RELATED TOPICS: Associative property of addition,
Commutative property of addition, Division**

The product of a number and zero is zero. This is also called the **zero property of multiplication.**

$$n \times 0 = 0, \quad 0 \times n = 0, \quad 4 \times 0 = 0, \quad 0 \times 4 = 0$$

Multiplying any number by 1 results in the same number. One is called the **multiplicative identity.**

$$n \times 1 = n \quad 1 \times n = n$$

● **RELATED TOPICS:** Zero property of addition, Additive identity

How to

Multiply factors of two or more digits:

1. Arrange the factors one under the other.
2. Multiply each digit in the multiplicand by each digit in the multiplier. The product of the multiplicand and each digit in the multiplier gives a **partial product.**
   a. To start, multiply the ones digit in the multiplier by the multiplicand from right to left.
   b. Align each partial product with its rightmost digit directly under its multiplier digit.
3. Add the partial products.

EXAMPLE   To multiply 204 × 103:

$$
\begin{array}{r}
\overset{1}{2}04 \\
\times \quad 103 \\
\hline
612 \\
0\ 00 \\
20\ 4 \\
\hline
21{,}012
\end{array}
$$

Multiply: 3 × 204 = 612. Align 612 under 3 in the multiplier.
Multiply: 0 × 204 = 000. Align 000 under 0 in the multiplier.
Multiply: 1 × 204 = 204. Align 204 under 1 in the multiplier.
Add the partial products as they are aligned.

Partial products 000 and 204 could be combined on a single line.

$$
\begin{array}{r}
204 \\
\times \quad 103 \\
\hline
612 \\
20\ 40 \\
\hline
21{,}012
\end{array}
$$

0 × 204 = 0. Align under 0 of the multiplier.
1 × 204 = 204. Align 204 under 1 of the multiplier on the same line.

● **RELATED TOPICS:** Area of rectangle, Multiplying integers

A shortcut can be used to multiply numbers that end with zero.

How to

Multiply factors that have ending zeros:

1. Separate the ending zeros from the other digits.
2. Multiply the remaining digits.
3. Attach the total number of ending zeros from the two factors to the product from Step 2.

EXAMPLE To multiply 2,600 × 70:

1.
$$\begin{array}{r} 26\!\mid\!00 \\ \times\ \ 7\!\mid\!0 \\ \hline \end{array}$$
Separate the ending zeros from the other digits.

2.
$$\begin{array}{r} 26\!\mid\!00 \\ \times\ \ 7\!\mid\!0 \\ \hline 182\ \ \end{array}$$
Multiply the other digits as if the zeros were not there. (26 × 7 = 182)

3.
$$\begin{array}{r} 26\!\mid\!00 \\ \times\ \ 7\!\mid\!0 \\ \hline 182000 \end{array}$$
Attach the zeros to the basic product. Note that the number of zeros affixed to the basic product is now the same as the sum of the number of zeros at the end of each factor.

● **RELATED TOPIC: Multiplying by powers of 10**

How to

Estimate the answer for a multiplication problem:

1. Round both factors to a chosen or specified place value or to one nonzero digit.
2. Multiply the rounded numbers.

How to

Check a multiplication problem:

1. Interchange the factors if convenient. Multiply the numbers a second time.
2. Compare the two products.

EXAMPLE   To multiply 284 × 41, first round and estimate. The exact product is then found. We can also check the answer and compare the three answers for reasonableness.

	Estimate	Exact	Check
2 8 4	3̣00	284	41
× 4̣ 1	4̣0	× 41	× 284
	12̣000	284	164
		11 36	3 28
		11,644	8 2
			11,644

The two exact answers are reasonably close to the estimate. The product is 11,644.

L_

The **distributive property of multiplication** means that multiplying a sum or difference by a factor is equivalent to multiplying each term of the sum or difference by the factor.

How to

Apply the distributive property of multiplication:

1. Add or subtract the numbers within the grouping.
2. Multiply the result of Step 1 by the factor outside the grouping.

or

1a. Multiply each number inside the grouping by the factor outside the grouping.
2a. Add or subtract the products from Step 1a.

Symbolically,

$$a \times (b + c) = a \times b + a \times c \quad \text{or} \quad a(b + c) = ab + ac$$
$$a \times (b - c) = a \times b - a \times c \quad \text{or} \quad a(b - c) = ab - ac$$

TIP

Other notations for multiplication

- Parentheses show multiplication when the distributive property is used: $a(b + c)$ means $a \times (b + c)$.
- The letters represent numbers.
- Letters written together with no operation sign between them imply multiplication: ab means $a \times b$; ac means $a \times c$.

EXAMPLE   According to the distributive property, 3(2 + 4) can be worked two ways:

Multiplying first gives: Adding first gives:

$$3\,(2 + 4) =$$ $$3\,(2 + 4) =$$
$$3\,(2) + 3\,(4) =$$ $$3\,(6) = 18$$
$$6 + 12 = 18$$

└

● **RELATED TOPICS:** Order of operations, Perimeter of rectangle

---

Division is the **inverse operation** of multiplication. Since 4 × 7 = 28, then 28 divided by 7 is 4 and 28 divided by 4 is 7. The number being divided is called the **dividend**. The number divided by is the **divisor**. The result is the **quotient**.

Division is *not* commutative. 12 ÷ 6 = 2, but 6 ÷ 12 does not equal 2; that is 6 ÷ 12 ≠ 2.

Division is *not* associative. (12 ÷ 6) ÷ 2 = 2 ÷ 2 = 1. 12 ÷ (6 ÷ 2) = 12 ÷ 3 = 4. That is, (12 ÷ 6) ÷ 2 ≠ 12 ÷ (6 ÷ 2).

● **RELATED TOPICS: Associative property of multiplication, Commutative property of multiplication**

How to

Write division symbolically:

1. To use the **divided by** symbol (÷), write the dividend first.

$$28 ÷ 7 = 4 \longleftarrow \text{ quotient}$$
dividend ⎯⎯↑   ↑⎯⎯ divisor

2. To use the **long-division symbol** (|‾), write the dividend under the bar.

$$4 \longleftarrow \text{quotient}$$
$$7\overline{)28}$$
divisor ⎯⎯↑ ↑⎯⎯ dividend

3. To use the **division bar** symbol or **slash**, write the dividend on top or first.

28 ⎯⎯⎯⎯ dividend
⎯ = 4 ⎯⎯ quotient   or   28/7 = 4 ⎯⎯ quotient
7 ⎯⎯⎯⎯ divisor

Which number goes first when dividing?

When using the "divided by" symbol, the division bar, or the slash as division, the dividend is written first or on top. The divisor is written second or on bottom.

Technology tools such as the calculator and computer normally use the "divided by" or slash symbol.

The divisor is written first *only* when using the long division symbol.

When the quotient is not a whole number, the quotient may have a **whole number part** and a **remainder.** When a dividend has more digits than a divisor, parts of the dividend are called **partial dividends,** and the quotient of a partial dividend and the divisor is called a **partial quotient.**

How to

Divide whole numbers:

1. Beginning with its leftmost digit, identify the first group of digits of the dividend that is larger than or equal to the divisor. This group of digits is the first *partial dividend.*
2. For each partial dividend in turn, beginning with the first:
   a. Divide the partial dividend by the divisor. Write this partial quotient above the rightmost digit of the partial dividend.
   b. Multiply the partial quotient by the divisor. Write the product below the partial dividend, aligning places.
   c. Subtract the product from the partial dividend. Write the difference below the product, aligning places. The difference must be less than the divisor.
   d. Next to the ones place of the difference, write the next digit of the dividend. This is the new partial dividend.
3. When all the digits of the dividend have been used, write the final difference in Step 2c as the remainder (unless the remainder is 0). The whole-number part of the quotient is the number written above the dividend.

EXAMPLE   To divide 881 by 35:

$35\overline{)88\,1}$ The first partial dividend is 88.

$$\begin{array}{r} 2 \\ 35\overline{)88\,1} \\ 70 \\ \hline 18 \end{array}$$
The partial quotient for 88 ÷ 35 is 2. Multiply 2 × 35 = 70. Then subtract 88 − 70 = 18. The difference 18 is less than the divisor 35.

$$\begin{array}{r} 2 \\ 35\overline{)881} \\ 70 \\ \hline 181 \end{array}$$
One from the dividend is written next to 18 to form the next partial dividend.

$$\begin{array}{r} 2\,5 \\ 35\overline{)881} \\ 70 \\ \hline 181 \\ 175 \\ \hline 6 \end{array}$$
The partial quotient for 181 ÷ 35 is 5. The product of 5 × 35 is 175. The difference of 181 − 175 is 6. The remainder is 6.

881 divided by 35 = **25 R6.**

● **RELATED TOPICS:** Dividing decimals, Dividing integers

Importance of placing the first digit carefully

The correct placement of the first digit in the quotient is critical. If the first digit is out of place, all digits that follow will be out of place, giving the quotient too few or too many digits.

EXAMPLE   To divide $5\overline{)2{,}535}$.

$$\begin{array}{r} 5 \\ 5\overline{)2{,}535} \\ 2\,5 \\ \hline 03 \end{array}$$
5 divides into 25 five times. Write 5 over the last digit of the 25. Subtract and bring down the 3.

$$\begin{array}{r} 5\,0\,7 \\ 5\overline{)2{,}535} \\ 2\,5 \\ \hline 035 \\ 35 \\ \hline 0 \end{array}$$
5 divides into 3 zero times. Write 0 over the 3 of the dividend. Bring down the next digit, which is 5. Divide 5 into the 35: 35 ÷ 5 = 7. Write the 7 over the 5 of the dividend. Multiply 7 × 5 = 35. Subtract.

The quotient of zero divided by a nonzero number is zero:

$$0 \div n = 0, \quad n\overline{)0}, \quad \frac{0}{n} = 0$$

$$0 \div 5 = 0, \quad 5\overline{)0}, \quad \frac{0}{5} = 0$$

The quotient of a number divided by zero is **undefined** or **indeterminant**:

$$n \div 0 \text{ is } undefined, \qquad 0\overline{)n}, \qquad \frac{n}{0} \text{ is undefined}$$

12 ÷ 0 is *undefined*

0 ÷ 0 is *indeterminant*

Dividing any nonzero number by itself yields 1:

$n \div n = 1$,   if n is not equal to zero; $12 \div 12 = 1$

Dividing any number by 1 yields the same number:

$n \div 1 = n$, $5 \div 1 = 5$

What types of situations require division?

Two types of common situations require division. Both types involve distributing items equally into groups.

1. Distribute a specified total quantity of items so that each group gets a specific equal share. Division determines the number of groups.

   For example, you need to ship 75 crystal vases. With appropriate packaging to avoid breakage, only 5 vases fit in each box. How many boxes are required? You divide the total quantity of vases by the quantity of vases that will fit into one box to determine how many boxes are required.

   75 crystal vases ÷ 5 vases per box = 15 boxes needed

2. Distribute a specified total quantity so that we have a specific number of groups. Division determines each group's equal share.

   For example, how many ounces will each of four cups contain if a carafe of coffee containing 20 ounces is poured equally into the cups? The capacity of the carafe is divided by the number of coffee cups:

   20 ounces ÷ 4 coffee cups = 5 ounces for each cup

How to

Estimate division:

1. Round the divisor and dividend to one nonzero digit.
2. Find the first digit of the quotient.
3. Attach a zero in the quotient for each remaining digit in the dividend.

How to

Check division:

1. Multiply the quotient by the divisor.
2. Add any remainder to the product in Step 1.
3. The result of Step 2 should equal the dividend.

EXAMPLE To estimate, to find the exact answer, and to check $913 \div 22$:

Estimate:
$$\begin{array}{r} 4\,0 \\ 20\,\overline{)90\,0} \end{array}$$
20 divides into 90 four whole times.
Attach a zero after 4.

Exact:
$$\begin{array}{r} 41\text{R}11 \\ 22\,\overline{)913} \\ 88 \\ \overline{33} \\ 22 \\ \overline{11} \end{array}$$

Check:
$$\begin{array}{r} 41 \\ \times\ 22 \\ \overline{82} \\ 82 \\ \overline{902} \end{array} \qquad \begin{array}{r} 902 \\ +\ 11 \\ \overline{913} \end{array}$$

The answer checks.

● **RELATED TOPICS:** Average, Dividing integers, Mean

■ **1–3 Exponents, Roots, and Powers of 10**

The product of repeated factors can be written in shorter form using natural-number exponents. **Natural numbers** are also called **counting numbers** and include all whole numbers except zero. Exponents that are not natural will have a different interpretation. For example, $4 \times 4 \times 4 = 4^3$. The 4 is the **base** and is the repeated factor. The 3 is the **exponent** and indicates the number of times the factor is repeated. The expression 4^3 is read "four cubed" or "four raised to the third **power.**" The expression 4^3 is written in **exponential notation.** The result of 4^3 is 64. The number 64 is written in **standard notation** and is called the **power.**

$$\text{base} \longrightarrow 4^{\overset{\displaystyle \text{exponent}}{3}} = 64 \longleftarrow \text{power}$$

How to

Change from exponential notation to standard notation:

1. Use the base as a factor as many times as indicated by the exponent.
2. Perform the multiplication.

EXAMPLE   To identify the base and exponent of the expression 5^3 and to write in standard notation:

5^3	5 is the base; 3 is the exponent.
$5^3 = 5 \times 5 \times 5$	Use the base as a factor 3 times. Perform the multiplication.
$= 125$	Standard notation.

Any number with an exponent of 1 is the number itself:

$$a^1 = a \text{ for any base } a, \qquad 8^1 = 8, \qquad 10^1 = 10$$

A number written without an exponent assumes an exponent of 1.

Any number (except zero) written with an exponent of 0 equals 1. The expression 0^0 is indeterminant.

$$a^0 = 1 \text{ for any nonzero base } a, \qquad 6^0 = 1, \qquad 27^0 = 1$$

● **RELATED TOPICS: Laws of exponents, Negative exponents, Powers of integers, Rational exponents**

The result of using a number as a factor 2 times is a square number, or a **perfect square.** In the expression $7^2 = 49$, the 49 is a perfect square.

The inverse operation of squaring is taking the **square root** of a number. The **principal square root** of a perfect square is the number that is used as a factor twice to equal that perfect square. The principal square root of 9 is 3 because 3^2 or $3 \times 3 = 9$.

The **radical sign** $\sqrt{}$ indicates that the square root is to be taken of the number under the bar. This bar serves as a grouping symbol just like parentheses. The number under the bar is called the **radicand**. The entire expression is called a **radical expression**.

radical sign ⟶ ↓ ↓ ⟵ bar

$$\sqrt{25} = 5 \longleftarrow \text{principal square root}$$

radicand ⟶

● **RELATED TOPICS:** Fractional exponents, Integers, Irrational numbers, Roots

How to

Find the square root of a perfect square by estimation:

1. Select a trial estimate of the square root.
2. Square the estimate.
3. If the square of the estimate is less than the original number, adjust the estimate to a larger number. If the square of the estimate is more than the original number, adjust the estimate to a smaller number.
4. Square the adjusted estimate from Step 3.
5. Continue the adjusting process until the square of the trial estimate is the original number.

EXAMPLE To find $\sqrt{256}$:

Select 15 as the estimated square root: 15^2 or $15 \times 15 = 225$. The number 225 is less than 256, so the square root of 256 must be larger than 15. We adjust the estimate to 17: 17^2 or $17 \times 17 = 289$. The number 289 is more than 256, so the square root of 256 must be smaller than 17. Now adjust the estimate to 16.

Because $16^2 = 256$, 16 is the square root of 256.

⌐

The squares of the numbers 1 through 10 are 1, 4, 9, 16, 25, 36, 49, 64, 81, 100. These are the only numbers from 1 through 100 that are perfect squares. We can expand the list whenever we need to. The more we use certain facts, the longer our memory retains them.

$11 \times 11 = 121$; $12 \times 12 = 144$; $13 \times 13 = 169$; $14 \times 14 = 196$;

$15 \times 15 = 225$; $16 \times 16 = 256$; $17 \times 17 = 289$; $18 \times 18 = 324$;

$19 \times 19 = 361$; $20 \times 20 = 400$

■ 1–4 Order of Operations

How to

Apply the order of operations:

1. **Parentheses (grouping symbols):** Perform operations within parentheses (or other grouping symbols), beginning with the innermost set of parentheses; or, apply the distributive property.
2. **Exponents and roots:** Evaluate exponential operations and find square roots in order from left to right.
3. **Multiply and divide** in order from left to right.
4. **Add and subtract** in order from left to right.

To summarize, use the following key words:

P arentheses (grouping), **E** xponents (and roots),
M ultiplication and **D** ivision, **A** ddition and **S** ubtraction

Other grouping symbols are **brackets [ ]**, **braces { }**, and a **bar.** The bar can combine with other symbols like the radical sign and be used as a grouping $\sqrt{4 + 5} = \sqrt{9} = 3$.

A memory aid for the order of operations

To remember the order of operations, use the sentence, " **P** lease **E** xcuse **M** y **D** ear **A** unt **S** ally."

- ■ **P** arentheses (grouping) ■ **E** xponents (roots)
- ■ **M** ultiply/ **D** ivide ■ **A** dd/ **S** ubtract

EXAMPLE   To simplify $4^2 - 5(2) \div (4 + 6)$:

$4^2 - 5(2) \div (4 + 6)$	Do operation within parentheses first: $4 + 6 = 10$.	P
$4^2 - 5(2) \div 10$	Evaluate exponentiation: $4^2 = 16$.	E
$16 - 5(2) \div 10$	Multiply: $5(2) = 10$.	M D
$16 - 10 \div 10$	Divide: $10 \div 10 = 1$.	M D
$16 - 1$	Subtract last.	A S
$16 - 1 = 15$		

Parentheses indicate multiplication or a grouping

Parentheses can indicate multiplication or an operation that should be done first. If the parentheses contain an operation, they indicate a grouping. Otherwise, they indicate multiplication. The expression 5(2) indicates multiplication, while (4 + 6) indicates a grouping.

● **RELATED TOPIC: Distributive property**

EXAMPLE   To evaluate $5 \times \sqrt{16} - 5 + [15 - (3 \times 2)]$:

$5 \times \sqrt{16} - 5 + [15 - (3 \times 2)]$ Work innermost grouping: 3×2.   P

$5 \times \sqrt{16} - 5 + [15 - 6]$ Work remaining grouping: $15 - 6$.   P

$5 \times \sqrt{16} - 5 + 9$ Find square root: $\sqrt{16} = 4$.   E

$5 \times 4 - 5 + 9$ Multiply: 5×4.   MD

$20 - 5 + 9$ Add and subtract from left to right.   AS

$15 + 9 = 24$

● **RELATED TOPIC: Order of operations for integers and signed numbers**

CHAPTER 2

Fractions

- 2–1 Concepts of Fractions
- 2–2 Operations with Fractions

■ 2–1 Concepts of Fractions

A **fraction** is a value that can be expressed as the quotient of two integers. A **common fraction** consists of two whole numbers. The bottom number, the **denominator,** indicates the number of parts one whole unit has been divided into. The top number, the **numerator,** tells how many of these parts are being considered. Fractions that represent less than one unit (less than 1), for example $\frac{1}{4}$ or $\frac{3}{4}$, are called **proper fractions.** Fractions that represent one or more units, for example $\frac{6}{6}$ or $\frac{7}{5}$, are called **improper fractions.**

A fraction is also called a **rational number.** An improper fraction that can be written as a combination of a whole number and a fractional part, for example $\frac{9}{4} = 2\frac{1}{4}$, is called a **mixed number.** A **complex fraction** has a fraction or mixed number in its numerator or denominator or both, for example $\frac{2\frac{1}{2}}{\frac{3}{4}}$. A **decimal fraction** is a fractional notation that has a denominator of 10 or some power of 10. In decimal notation the decimal point and the place values to its right represent the denominator of the decimal fraction. A **mixed decimal fraction** is a notation used to represent a mixed number that contains a whole number part and a decimal fractional part.

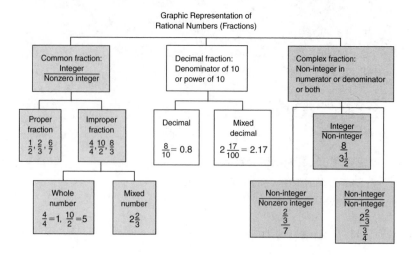

Graphic Representation of
Rational Numbers (Fractions)

● **RELATED TOPICS: Decimals, Integers**

The **natural numbers** are the **counting numbers**—1, 2, 3, 4, 5, 6, 7, 8, and so forth. A **multiple** of a natural number is the product of that number and another natural number. Multiples of 2 are 2, 4, 6, 8, 10, etc.

How to

Find multiples of a natural number:

1. Write the product of the given natural number and 1.
2. Write the product of the given natural number and 2.
3. Write the product of the given natural number and 3.
4. Continue in the same manner.

EXAMPLE   To show that 3, 6, and 9 are multiples of 3:

$3 = \boxed{3} \times 1$ $\qquad$ $6 = \boxed{3} \times 2$ $\qquad$ $9 = \boxed{3} \times 3$ Write each number as the product of 3 and a natural number.

Zero and natural numbers that are multiples of 2 are **even numbers**.
Natural numbers that are not multiples of 2 are **odd numbers**.
A number is **divisible** by another number if the quotient has no remainder or if the dividend is a multiple of the divisor.

How to

Test for divisibility:

A number is divisible by

1. 2 if the last digit is an even number (0, 2, 4, 6, or 8).
2. 3 if the sum of its digits is divisible by 3.
3. 4 if the last two digits form a number that is divisible by 4.
4. 5 if the last digit is 0 or 5.
5. 6 if the number is divisible by *both* 2 and 3.
6. 7 if the division has no remainder.
7. 8 if the last three digits form a number divisible by 8.
8. 9 if the sum of its digits is divisible by 9.
9. 10 if the last digit is 0.

EXAMPLE   To identify the number in each pair that is divisible by the given divisor:

Numbers	Divisor	Answer
874 or 873	2	**874;** the last digit is an even digit (4).
427 or 423	3	**423;** the sum of the digits is divisible by 3: $4 + 2 + 3 = 9$.
5,912 or 5, 913	4	**5,912;** the last two digits form a number divisible by 4: $12 \div 4 = 3$.
80 or 82	5	**80;** the last digit is 0.
804 or 802	6	**804;** the last digit is even and the sum of the digits is divisible by 3.
477 or 475	9	**477;** the sum of the digits is divisible by 9: $4 + 7 + 7 = 18$.
182 or 180	10	**180;** the last digit is 0.

● **RELATED TOPIC: Dividing whole numbers**

A **factor pair** of a natural number is a pair of natural numbers that has a product equal to the given natural number. Every natural number greater than 1 has at least one factor pair, 1 and the number itself. The factor pair of 3 is 1×3. Many natural numbers have more than one factor pair. The factor pairs of 12 are 1×12, 2×6, and 3×4.

How to

Find all factor pairs of a natural number:

1. Write the factor pair of 1 and the number.
2. Check to see if the number is divisible by 2. If so, write the factor pair of 2 and the quotient of the beginning natural number and 2.
3. Check each natural number for divisibility until you reach a number that has already appeared as a quotient in a previous factor pair.

EXAMPLE   To list all the factor pairs of 18:

1×18	Start with 1×18.
2×9	18 is divisible by 2: $18 \div 2 = 9$.
3×6	18 is divisible by 3: $18 \div 3 = 6$. 18 is not divisible by 4 or 5. 18 is divisible by 6; 6 is in the factor pair 3×6, so we stop.

Factor pairs of 18 are 1 and 18, 2 and 9, and 3 and 6.

How to

Find all factors of a natural number:

1. List all factor pairs of the number.
2. Arrange each distinct factor in order from smallest to largest.

EXAMPLE   To list all factors of 18:

1×18	Factor pairs of 18
2×9	
3×6	

Factors of 18: 1, 2, 3, 6, 9, 18

● **RELATED TOPIC: Factoring polynomials**

A **prime number** is a whole number greater than 1 that has only one factor pair, the number itself and 1. Note that 1 is not a prime number. A **composite number** is a whole number greater than 1 that is not a prime number.

EXAMPLE   To identify the composite and prime numbers by examining factor pairs:

  (a)   4   (b)   10   (c)   13   (d)   12   (e)   5

  (a)   **4 is a composite number** because its factor pairs are 1×4 and 2×2.

  (b)   **10 is a composite number** because its factor pairs are 1×10 and 2×5.

  (c)   **13 is a prime number** because its only factor pair is 1×13.

  (d)   **12 is a composite number** because its factor pairs are 1×12, 2×6, and 3×4.

  (e)   **5 is a prime number** because its only factor pair is 1×5.

● **RELATED TOPICS: Prime factorization; Prime polynomials**

Prime numbers less than 50

We can find all the prime numbers that are 50 or less using an ancient technique developed by the mathematician Erastosthenes.

  1. List the numbers from 1 through 50.
  2. Eliminate numbers that are not prime using the systematic process:
      (a)   1 is not prime. Eliminate 1.
      (b)   2 is prime. Eliminate all multiples of 2.
      (c)   3 is prime. Eliminate all multiples of 3.
      (d)   4 has already been eliminated.
      (e)   5 is prime. Eliminate all multiples of 5.
      (f)   6 has already been eliminated.
      (g)   7 is prime. Eliminate all multiples of 7.
  3. Circle remaining numbers as prime numbers.

1̸	②	③	4̸	⑤	6̸	⑦	8̸	9̸	1̸0̸
⑪	1̸2̸	⑬	1̸4̸	1̸5̸	1̸6̸	⑰	1̸8̸	⑲	2̸0̸
2̸1̸	2̸2̸	㉓	2̸4̸	2̸5̸	2̸6̸	2̸7̸	2̸8̸	㉙	3̸0̸
㉛	3̸2̸	3̸3̸	3̸4̸	3̸5̸	3̸6̸	㊲	3̸8̸	3̸9̸	4̸0̸
㊶	4̸2̸	㊸	4̸4̸	4̸5̸	4̸6̸	㊼	4̸8̸	4̸9̸	5̸0̸

All numbers not already eliminated are prime. Why? The numbers 8, 9, and 10 have already been eliminated as multiples of 2, 3, and 5, respectively. Multiples of 11 that are less than 50 have already been eliminated: $11 \times 2 = 22$, $11 \times 3 = 33$, $11 \times 4 = 44$. $11 \times 5 = 55$ is greater than 50. Similarly, all other composite numbers have already been eliminated.

Prime factorization refers to writing a composite number as the product of only prime numbers. Factors of a number that are prime numbers are called **prime factors.**

How to

Find the prime factors of a composite number:

1. Test the composite number to determine if it is divisible by a prime. Start with 2.
2. Make a factor pair using the first prime number that passes the test in Step 1.
3. Carry forward the prime factors and test the remaining composite factor by repeating Steps 1 and 2.
4. Continue until all factors are prime. Repeated prime factors can be written in exponential notation.

EXAMPLE   To write the prime factorization of 30:

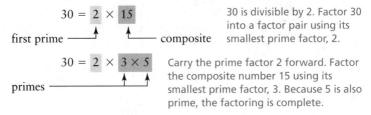

$30 = 2 \times 15$

first prime ——— composite

30 is divisible by 2. Factor 30 into a factor pair using its smallest prime factor, 2.

$30 = 2 \times 3 \times 5$

primes ————

Carry the prime factor 2 forward. Factor the composite number 15 using its smallest prime factor, 3. Because 5 is also prime, the factoring is complete.

The prime factorization of 30 is 2 × 3 × 5.

EXAMPLE   To write the prime factorization of 18:

$18 = 2 \times 9$ 18 is divisible by 2. Factor 18 into a factor pair using its smallest prime factor, 2. Factor 9 using its smallest prime factor, 3.

$18 = 2 \times 3 \times 3$ $9 = 3 \times 3$. The number 3 is prime so factoring is complete.

$18 = 2 \times 3^2$ Exponential notation.

● **RELATED TOPIC: Exponential notation**

The **least common multiple (LCM)** of two or more natural numbers is the smallest number that is a multiple of each number. The LCM is divisible by each number.

How to

Find the least common multiple of two numbers by inspection:

1. List the first few multiples of each number.
2. Examine each set of multiples to find the *smallest* number that is a multiple for both numbers.

EXAMPLE   To write the least common multiple of 3 and 5:
Write the first few multiples of each number, then find the smallest number that is a multiple of both numbers.

Multiples of 3: 3, 6, 9, 12, 15 , 18, 21, 24, 27, 30 , 33, ...

Multiples of 5: 5, 10, 15 , 20, 25, 30 , 35, ...

Both 15 and 30 are common multiples, but **15 is the least common multiple.**

How to

Find the least common multiple of numbers by using the prime factorization of the numbers:

1. List the prime factorization of each number using exponential notation.
2. List the prime factorization of the least common multiple by including the prime factors appearing in *each* number. If a prime factor appears in more than one number, use the factor with the *largest* exponent.
3. Write the resulting expression in standard notation.

EXAMPLE   To find the least common multiple of 12 and 40 by prime factorization:

$12 = 2 \times 2 \times 3 \qquad = 2^2 \times 3$ Prime factorization of 12.

$40 = 2 \times 2 \times 2 \times 5 = 2^3 \times 5$ Prime factorization of 40.

$\text{LCM} = 2^3 \times 3 \times 5$ Prime factorization of LCM.

$\textbf{LCM} = \textbf{120}$ LCM in standard notation.

● **RELATED TOPIC:** Least common denominator (LCD)

The **greatest common factor (GCF)** of two or more numbers is the largest factor common to each number. Each number is divisible by the GCF.

How to

Find the greatest common factor (GCF) of two or more natural numbers:

1. List the prime factorization of each number using exponential notation when appropriate.
2. List the prime factorization of the greatest common factor by including each prime factor appearing in *every* number. If a prime factor appears more than one time in a number, use the factor with the *smallest* exponent. If there are no common prime factors, the GCF is 1.
3. Write the resulting expression in standard notation.

EXAMPLE   To find the greatest common factor of 15, 30, and 45.

$15 = 3 \times 5 \qquad = 3 \times 5$ Prime factorization of 15.

$30 = 2 \times 3 \times 5 = 2 \times 3 \times 5$ Prime factorization of 30.

$45 = 3 \times 3 \times 5 = 3^2 \times 5$ Prime factorization of 45.

$\text{GCF} = 3 \times 5$ Common prime factors.

$\textbf{GCF} = \textbf{15}$ GCF in standard notation.

● **RELATED TOPIC:** Factor polynomials

Equivalent fractions have the same value. A fraction is in **lowest terms** if no whole number divides evenly into *both* the numerator and denominator except the number 1.

The **Fundamental Principle of Fractions** states that if the numerator and denominator of a fraction are multiplied by the same nonzero number, the value of the fraction remains unchanged.

How to

Write an equivalent fraction with a higher denominator:

1. Multiply the numerator by any rational number.
2. Multiply the denominator by the same number used in Step 1.

EXAMPLE   To find three fractions that are equivalent to $\frac{1}{2}$:

$$\frac{1}{2} \times \boxed{\frac{2}{2}} = \frac{2}{4}$$ Multiply by 1 in the form of $\frac{2}{2}$.

$$\frac{1}{2} \times \boxed{\frac{3}{3}} = \frac{3}{6}$$ Multiply by 1 in the form of $\frac{3}{3}$.

$$\frac{1}{2} \times \boxed{\frac{4}{4}} = \frac{4}{8}$$ Multiply by 1 in the form of $\frac{4}{4}$.

Multiplication, division, and 1

In the preceding example $\frac{1}{2}$ is multiplied by a fraction that has a value of 1, and 1 times any number does not change the value of that number. Written symbolically,

$$\frac{n}{n} = 1 \quad \text{and} \quad 1 \times n = n, n \neq 0$$

Fractions in the same family can be generated by multiplying the fraction by 1 in the form of $\frac{2}{2}, \frac{3}{3}, \frac{4}{4}$, and so on.

How to

Change a fraction to an equivalent fraction with a specified higher denominator:

1. Divide the larger denominator by the original denominator.
2. Multiply the original numerator and denominator by the quotient found in Step 1.

EXAMPLE   To change $\frac{5}{8}$ to an equivalent fraction that has a denominator of 32:

$$\frac{5}{8} = \frac{?}{32}, \qquad 32 \div 8 = \boxed{4} \quad \text{Multiply by 1 in the form of } \frac{4}{4}.$$

$$\frac{5}{8} \times \frac{4}{4} = \frac{20}{32}$$

● **RELATED TOPIC: Equivalent rational expressions**

How to

Change a fraction to an equivalent fraction with a smaller denominator or reduce a fraction to lowest terms:

1. Find a common factor greater than 1 for the numerator and denominator.
2. Divide both the numerator and denominator by this common factor.
3. Continue until the fraction is in lowest terms or has the desired smaller denominator.

Note: To **reduce to lowest terms** in the fewest steps, find the greatest common factor (GCF) in Step 1.

Reducing and the properties of 1

To reduce the fraction $\frac{8}{10}$, we divide by the whole number 1 in the form of $\frac{2}{2}$. $\frac{8}{10} \div \frac{2}{2} = \frac{4}{5}$. A nonzero number divided by itself is 1, and to divide a number by 1 does not change the value of the number. Symbolically,

$$\frac{n}{n} = 1; n \neq 0 \quad \text{and} \quad n \div 1 = n \quad \text{or} \quad \frac{n}{1} = n$$

EXAMPLE   To reduce $\frac{18}{24}$ to lowest terms:

Prime factors of 18: $2 \times 3 \times 3$ or 2×3^2

Prime factors of 24: $2 \times 2 \times 2 \times 3$ or $2^3 \times 3$

The GCF is 2×3 or 6.   Divide by 1 in the form of $\frac{6}{6}$.

$$\frac{18}{24} \div \frac{6}{6} = \frac{3}{4}$$

Do you have to use the GCF to reduce to lowest terms?

A fraction can be reduced to lowest terms in the fewest steps by using the *greatest common factor*; however, it can still be reduced using any common factor; this takes a few more steps.

$$\frac{18 \div 2}{24 \div 2} = \frac{9}{12} \qquad \frac{9 \div 3}{12 \div 3} = \frac{3}{4}$$

● **RELATED TOPIC: Reducing rational expressions**

How to

Write an improper fraction as a whole or mixed number:

Write $\frac{12}{3}$ and $\frac{13}{3}$ as whole or mixed numbers.

$$3\overline{\smash{)}12}\;\;\overset{4}{}\qquad 3\overline{\smash{)}13}\;\;\overset{4R\,1}{}\qquad \frac{12}{3}=4\qquad \frac{13}{3}=4\frac{1}{3}$$

1. Divide the numerator of the improper fraction by the denominator.
2. Examine the remainder.
   a. If the remainder is 0, the quotient is a whole number: the improper fraction is equivalent to this whole number.
   b. If the remainder is not 0, the quotient is not a whole number: the improper fraction is equivalent to a mixed number. The whole number part of this mixed number is the whole number part of the quotient. The numerator of the fraction part of the mixed number is the remainder; the denominator is the divisor (also the denominator of the improper fraction).

EXAMPLE   To write $\frac{139}{8}$ as a whole or mixed number:

$$8\overline{\smash{)}139}\;\;\overset{17\ R\ 3}{}, \text{ or } 17\frac{3}{8}$$
$$\underline{8}$$
$$59$$
$$\underline{56}$$
$$3$$

Divide 139 by 8. The quotient is 17 R3, which equals $17\frac{3}{8}$.

$$\frac{139}{8}=17\frac{3}{8}$$

Converting to a whole or mixed number is different from reducing

Do not confuse converting an improper fraction to a whole or mixed number with reducing to lowest terms. An improper fraction is in lowest terms if its numerator and denominator have no common factor. Therefore, the improper fraction $\frac{10}{7}$ is in lowest terms. The improper fraction $\frac{10}{4}$ is not in lowest terms. It will reduce to $\frac{5}{2}$, which is in lowest terms.

Fractions

EXAMPLE To convert $\frac{28}{8}$ to a mixed number:

$$\frac{28}{8} = \frac{7}{2} \qquad 2\overline{)7} \;\begin{array}{c}3\\ \end{array} = 3\frac{1}{2} \qquad \text{Fraction is reduced before dividing.}$$

$$\frac{6}{1}$$

or

$$\frac{28}{8} \qquad 8\overline{)28}\;\begin{array}{c}3\\\end{array} = 3\frac{4}{8} = 3\frac{1}{2} \qquad \text{Fraction is reduced after dividing.}$$

$$\frac{24}{4}$$

How to

Convert a mixed number to an improper fraction:

1. Multiply the denominator of the fractional part by the whole number.
2. Add the numerator of the fractional part to the product; this sum becomes the numerator of the improper fraction.
3. The denominator of the improper fraction is the same as the denominator of the fractional part of the mixed number.

EXAMPLE To change $6\frac{2}{3}$ to an improper fraction:

$$6\frac{2}{3} = \frac{(3 \times 6) + 2}{3} = \frac{20}{3}$$

How to

Convert a whole number to an improper fraction:

1. Write the whole number as a fraction with a denominator of 1.
2. Change the whole-number fraction to a fraction having a higher denominator by multiplying numerator and denominator by the higher denominator.

EXAMPLE To change 8 to fifths:

$$8 = \frac{8}{1} \times \frac{5}{5} = \frac{40}{5}$$

How to

Find the least common denominator (LCD) of two or more fractions:

Find the least common multiple (LCM) of the denominators of the fractions.

EXAMPLE   To find the least common denominator for the fractions $\frac{5}{12}, \frac{4}{15}, \frac{3}{8}$:
Write the prime factorization of each denominator: then find the LCM.

$$\begin{array}{ccc}
\underline{12} & \underline{15} & \underline{8} \\
2 \times 6 & 3 \times 5 & 2 \times 4 \\
2 \times 2 \times 3 & & 2 \times 2 \times 2 \\
2^2 \times 3 & & 2^3
\end{array}$$

$$\begin{aligned}
\text{LCM or LCD} &= 2^3 \times 3 \times 5 \\
&= 8 \times 3 \times 5 \\
&= \mathbf{120}
\end{aligned}$$

TIP

Alternative procedure for finding the LCM or LCD

We can also find the LCM or LCD of several fractions by dividing duplicated factors and then, multiplying.

1. Arrange the denominators horizontally.
2. Divide by any prime factor that divides evenly into at least two denominators.
3. The LCM or LCD is the product of the prime and remaining factors.

Look at the denominators from the preceding example.

$$\begin{array}{cccl}
2 \mid 8 & 12 & 15 & \text{Divide by the factor 2.} \\
2 \mid 4 & 6 & 15 & \text{Divide by the factor 2.} \\
3 \mid 2 & 3 & 15 & \text{Divide by the factor 3.} \\
2 & 1 & 5 &
\end{array}$$

Prime factors: 2, 2, 3
Remaining factors: 2, 1, 5
$$\begin{aligned}
\text{LCM} &= 2 \cdot 2 \cdot 3 \cdot 2 \cdot 1 \cdot 5 \\
&= 120
\end{aligned}$$

● **RELATED TOPICS: Exponential notation, Least common multiple**

How to

Compare fractions:

1. Find the least common denominator (LCD) that is also the least common multiple (LCM).
2. Change each fraction to an equivalent fraction with the least common denominator (LCD) as its denominator.
3. Compare the numerators.

EXAMPLE   To compare $\frac{3}{4}$ to $\frac{7}{12}$:

4:    4, 8, **12**, 16   12 is a multiple of 4.

12:    **12**, 24, 36   12 is a multiple of 12. LCD = 12.

$\frac{3}{4} \times \frac{3}{3} = \frac{9}{12}$ Change $\frac{3}{4}$ to an equivalent fraction. Multiply by 1 in the form of $\frac{3}{3}$.

$\frac{7}{12} \quad = \frac{7}{12}$ No change needed.

$\frac{9}{12}$ is greater than $\frac{7}{12}$ Compare the numerators.

● **RELATED TOPIC: Comparing decimals**

▲ **APPLICATION: Selecting Tool Size**

How to

Select the appropriate-sized tool:

1. Determine the approximate size tool needed.
2. Determine the tool sizes available.
3. Use knowledge of comparing fractions to compare the exact sizes of the tools and make the appropriate selection.

EXAMPLE   Two drill bits have diameters of $\frac{3}{8}$ in. and $\frac{5}{16}$ in., respectively. To determine which drill bit makes the larger hole:
The least common denominator is 16.

$$\frac{3}{8} = \frac{6}{16} \qquad \frac{5}{16} = \frac{5}{16}$$ Change $\frac{3}{8}$ to an equivalent fraction with a denominator of 16.

Now that each fraction has been changed to an equivalent fraction with the same denominator, we compare the numerators: $\frac{6}{16}$ is larger than $\frac{5}{16}$ because 6 is larger than 5, so $\frac{3}{8}$, which is equivalent to $\frac{6}{16}$, is larger than $\frac{5}{16}$. **The drill bit with a $\frac{3}{8}$-in. diameter will drill the larger hole.**

■ 2–2 Operations with Fractions

How to

Add fractions:

1. If the denominators are not the same, find the least common denominator.
2. Change each fraction not already expressed in terms of the common denominator to an equivalent fraction with the common denominator.
3. Add the numerators only.
4. The common denominator is the denominator of the sum.
5. Reduce the sum to lowest terms and change improper fractions to whole or mixed numbers.

EXAMPLE   To find the sum of $\dfrac{3}{8} + \dfrac{1}{8}$:

Start with Step 3 of the addition procedure because the denominators are the same.

$$\frac{3}{8} + \frac{1}{8} =$$ Add the numerators. The common denominator 8 is the denominator of the sum.

$$\frac{4}{8} =$$ Reduce to lowest terms.

$$\frac{1}{2}$$

EXAMPLE   To add $\dfrac{5}{32} + \dfrac{3}{16} + \dfrac{7}{8}$:

8: 8, 16, 24, **32**, 40   The least common denominator is 32.

16: 16, **32**, 48

32: **32**, 64

$$\frac{5}{32} = \frac{5}{32}, \qquad \frac{3}{16} \times \frac{2}{2} = \frac{6}{32}, \qquad \frac{7}{8} \times \frac{4}{4} = \frac{28}{32}$$

Change each fraction to an equivalent fraction with a denominator of 32.

$$\frac{5}{32} + \frac{6}{32} + \frac{28}{32} = \frac{39}{32}$$

Add the numerators.

$$\frac{39}{32} = 1\frac{7}{32}$$

Change to a mixed number.

TIP

Find the least common denominator for multiples

The least common denominator (LCD) for denominators that are multiples of the same number is the largest denominator. Since 8, 16, and 32 are multiples of 8 and 32 is the largest, 32 is the LCD.

▲ **APPLICATION: Outside Diameter of Circular Object**

How to

Find the outside diameter:

1. Determine the inside diameter and thickness of wall (or insulation, pipe, tube, etc.).
2. Add the inside diameter and the thickness twice.

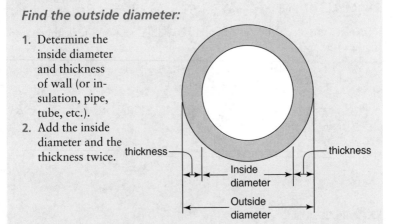

EXAMPLE   A plumber uses a $\frac{9}{16}$-in.-diameter copper tube wrapped with $\frac{5}{8}$-in. insulation. What size hole must he bore in the wall support to install the insulated pipe?

From Fig. 2–1, we see that the $\frac{5}{8}$-in. insulation increases the diameter on each side of the pipe. To get the total diameter of the pipe and insulation, we add $\frac{9}{16} + \frac{5}{8} + \frac{5}{8}$. The thickness of the insulation is added twice because it counts in the total diameter of the pipe and insulation two times.

$$\frac{9}{16} = \frac{9}{16} \qquad \text{The LCD is 16.}$$

$$\frac{5}{8} = \frac{10}{16}$$

$$+\ \frac{5}{8} = \frac{10}{16}$$

$$\frac{29}{16} = 1\frac{13}{16}$$

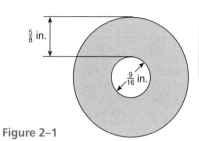

Figure 2–1

The total diameter is $1\frac{13}{16}$ in., and the diameter of the hole must be at least this large.

How to

Add mixed numbers:

1. Add the whole-number parts.
2. Add the fractional parts and reduce to lowest terms.
3. Change improper fractions to whole or mixed numbers.
4. Add whole-number parts.

EXAMPLE   To add $5\frac{2}{3} + 7\frac{3}{8} + 4\frac{1}{2}$.

$$5\frac{2}{3} = 5\frac{16}{24} \qquad \text{The LCD is 24. Change fractions to equivalent fractions.}$$

$$7\frac{3}{8} = 7\frac{9}{24} \qquad \text{Add whole numbers.}$$

$$+\ 4\frac{1}{2} = 4\frac{12}{24} \qquad \text{Add fractional parts.}$$

$$16\frac{37}{24} \qquad \frac{37}{24} = 1\frac{13}{24} \qquad \text{Change improper fraction to a mixed number.}$$

$$16 + 1\frac{13}{24} = 17\frac{13}{24} \qquad \text{Add whole-number parts.}$$

Adding whole numbers and mixed numbers

When adding a whole number and a mixed number, think of the whole number as a mixed number with zero as the numerator in the fraction.

$$5 + 3\frac{1}{3} = 5\frac{0}{3} + 3\frac{1}{3} \qquad \begin{array}{r} 5\frac{0}{3} \\[6pt] +\ 3\frac{1}{3} \\ \hline 8\frac{1}{3} \end{array}$$

● **RELATED TOPICS: Adding fractions, Adding whole numbers, Equivalent fractions, Improper fractions, Reducing fractions**

How to

Subtract fractions:

1. If the denominators are not the same, find the least common denominator.
2. Change each fraction not expressed in terms of the common denominator to an equivalent fraction having the common denominator.
3. Subtract the numerators.
4. The common denominator will be the denominator of the difference.
5. Reduce the difference to lowest terms.

EXAMPLE   Subtract $\dfrac{3}{8} - \dfrac{7}{32}$.

8 and 32 are both multiples of 8. Since 32 is larger, it is the LCD.

$$\frac{3}{8} = \frac{12}{32}$$ Change $\frac{3}{8}$ to an equivalent fraction with a denominator of 32.

$$-\frac{7}{32} = \frac{7}{32}$$ Subtract numerators and keep the common denominator.

$$\frac{5}{32}$$ The difference is written in lowest terms.

How to

Subtract mixed numbers:

1. If the fractional parts of the mixed numbers do not have the same denominator, change them to equivalent fractions with a common denominator.
2. When the fraction in the minuend is larger than the fraction in the subtrahend, go to Step 4.
3. When the fraction in the subtrahend is larger than the fraction in the minuend, borrow (regroup) one whole number from the whole-number part of the minuend. This makes the whole number 1 less.
4. Change the whole number borrowed to an improper fraction with the common denominator. For example, $1 = \frac{3}{3}$, $1 = \frac{8}{8}$, $1 = \frac{n}{n}$, where n is the common denominator.
5. Add the borrowed fraction ($\frac{n}{n}$) to the fraction already in the minuend.
6. Subtract the fractional parts and the whole-number parts.
7. Reduce the answer to lowest terms.

EXAMPLE   To subtract $15\frac{3}{4}$ from $18\frac{1}{2}$:

$$18\frac{1}{2} = 18\frac{2}{4} = 17\frac{4}{4} + \frac{2}{4} = 17\frac{6}{4}$$

Borrow 1 from 18.
$18 - 1 = 17$, $1 = \frac{4}{4}$, $\frac{4}{4} + \frac{2}{4} = \frac{6}{4}$

$$- 15\frac{3}{4} = 15\frac{3}{4} = 15\frac{3}{4} \qquad = 15\frac{3}{4}$$

Subtract fractions.
Subtract whole numbers.

$$2\frac{3}{4}$$

● **RELATED TOPICS:** Borrowing, Equivalent fractions, Subtracting fractions, Subtracting whole numbers

▲ **APPLICATION:** Tolerance and Limit Dimensions

The **tolerance** of a measurement is the specified amount a measurement can vary. The **maximum measurement** is the ideal measurement plus the tolerance. The **minimum measurement** is the ideal measurement minus the tolerance. The maximum and minimum measurements together are known as the **limit dimensions** or the acceptable interval of tolerance.

How to

Find the limit dimensions of a measurement:

1. To find the maximum limit, add the tolerance to the specified dimension.
2. To find the minimum limit, subtract the tolerance from the specified dimension.

EXAMPLE   To find the limit dimensions and the tolerance interval of a part if the blueprint calls for the part to be 2 in. long and the tolerance is $\pm\frac{1}{8}$ in. ($\pm$ is read "plus or minus"):

$$2 - \frac{1}{8} = 1\frac{8}{8} - \frac{1}{8} = 1\frac{7}{8} \qquad \text{Find the minimum.}$$

$$2 + \frac{1}{8} = 2\frac{1}{8} \qquad\qquad\qquad \text{Find the maximum.}$$

The limit dimensions are $1\frac{7}{8}$ in. and $2\frac{1}{8}$ in. The tolerance interval is from $1\frac{7}{8}$ to $2\frac{1}{8}$.

 TIP

Think of whole numbers in mixed-number form before subtracting

As in addition, when subtracting whole numbers and mixed numbers, consider the whole number to have zero fractional parts. Then follow the same procedures as before. Borrow when necessary.

EXAMPLE   To subtract 27 from $45\frac{1}{3}$:

$$45\frac{1}{3} = 45\frac{1}{3}$$

$$\underline{-\ 27 = 27\frac{0}{3}}$$
$$\qquad\qquad 18\frac{1}{3}$$

Write a zero fraction.
Subtract fractions.
Subtract whole numbers.

● **RELATED TOPICS: Subtracting fractions, Subtracting whole numbers**

EXAMPLE   To find the number of feet of wire left on a 100-ft roll if $27\frac{1}{4}$ ft are used from the roll:

$$100 \;=\; 99\frac{4}{4}$$ Borrow 1 from 100 and write as the fraction $\frac{4}{4}$.

$$\begin{array}{r} -\,27\frac{1}{4} = 27\frac{1}{4} \\ \hline 72\frac{3}{4} \end{array}$$ Subtract fractions.
Subtract whole numbers.

$72\frac{3}{4}$ ft of wire is left on the roll.

⌐

▲ **APPLICATION: Inside Diameter of Circular Object**

The **inside diameter** of a circular object is the measurement of the diameter taken from the inside of the pipe through the center to the opposite inside.

How to

Find the inside diameter of a circular object:

1. Determine the outside diameter and thickness of wall.
2. Add the wall thickness 2 times.
3. Subtract the result of Step 2 from the outside diameter.

EXAMPLE   The top of a circular planter has an outside diameter of 24 in. and a thickness of $4\frac{1}{2}$ in. To find the inside diameter of the top of the planter for a flower pot insert:

24 in.   outside diameter

$4\frac{1}{2}$ in.   thickness of wall

$$4\frac{1}{2} + 4\frac{1}{2} = 8\frac{2}{2} = 9$$ Multiply the wall thickness times 2.

$$24 - 9 =$$ Subtract wall thicknesses from the outside diameter.

15 in.   Inside diameter.

⌐

▲ APPLICATION: Missing Dimensions

How to

Find the missing dimension:

1. Add the known dimensions.
2. Subtract the sum of the known dimensions from the overall dimension.

EXAMPLE To find the missing length:

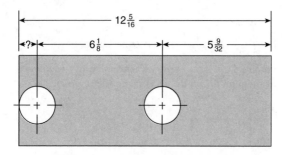

$$6\frac{1}{8} = 6\frac{4}{32} \qquad \text{Add the known parts.}$$

$$-5\frac{9}{32} = 5\frac{9}{32}$$

$$11\frac{13}{32}$$

$$12\frac{5}{16} = 12\frac{10}{32} \qquad \text{To subtract the sum of the known parts from the}$$
$$\qquad\qquad\qquad\quad \text{total length find a common denominator.}$$

$$-11\frac{13}{32} = 11\frac{13}{32}$$

$$12\frac{10}{32} = 11\frac{42}{32} \qquad \text{Borrow and rewrite as } \frac{32}{32} \text{ then add } \frac{32}{32} + \frac{10}{32} = \frac{42}{32}. \text{ Then}$$
$$\qquad\qquad\qquad\quad \text{subtract fractions. Subtract whole numbers.}$$

$$-11\frac{13}{32} = 11\frac{13}{32}$$

$$\frac{29}{32}$$

The missing length is $\frac{29}{32}$ in.

To find a *part of a part,* we multiply fractions.

$$\frac{1}{3} \times \frac{3}{5} = \frac{1}{5}$$

1 whole $\frac{3}{5}$ of 1 whole $\frac{1}{3}$ of $\frac{3}{5}$ of 1 whole is $\frac{1}{5}$ of 1 whole

Fractions

How to

Multiply fractions:

1. Reduce common factors in a numerator and denominator.
2. Multiply the numerators to get the numerator of the product.
3. Multiply the denominators to get the denominator of the product.
4. Reduce the product to lowest terms (not necessary if all common factors are reduced in step 1.)

TIP

Reduce or cancel before multiplying

To multiply fractions, reduce common factors:

$$\frac{1}{3} \times \frac{3}{5} = \frac{1 \times \overset{1}{\cancel{3}}}{\underset{1}{\cancel{3}} \times 5} = \frac{1}{5} \qquad \frac{1 \times 3}{3 \times 5} = \frac{1 \times 3}{5 \times 3} = \frac{1}{5} \times \frac{3}{3} = \frac{1}{5} \times 1 = \frac{1}{5}$$

In this example a numerator and a denominator both have a common factor of 3, so the common factor can be reduced before multiplying. **Reducing** applies the principles $\frac{n}{n} = 1$ and $n \times 1 = n$. This process is also referred to as **canceling**.

EXAMPLE   To find $\frac{3}{9}$ of $\frac{2}{7}$:

$$\frac{\overset{1}{\cancel{3}}}{\underset{3}{\cancel{9}}} \times \frac{2}{7} =$$ 3 is a common factor of both a numerator and a denominator. Reduce before multiplying.

$$\frac{1}{3} \times \frac{2}{7} =$$ Multiply numerators. Multiply denominators.

$$\frac{2}{21}$$ The fraction is in lowest terms.

Cancel from any numerator to any denominator in multiplication

Common factors that are reduced can be diagonal to each other, one above the other, or separated by another fraction, but one factor *must* be in the numerator and the other in the denominator.

$$\frac{2}{\underset{1}{3}} \times \frac{\overset{1}{3}}{5} = \frac{2}{5} \qquad \frac{\overset{3}{6}}{\underset{4}{8}} \times \frac{3}{5} = \frac{9}{20} \qquad \frac{\overset{1}{2}}{7} \times \frac{1}{3} \times \frac{5}{\underset{4}{8}} = \frac{5}{84}$$

How to

Multiply mixed numbers:

1. Change each mixed number or whole number to an improper fraction.
2. Reduce as much as possible.
3. Multiply numerators.
4. Multiply denominators.
5. Change the product to a whole or mixed number if possible.

EXAMPLE   To multiply $2\frac{1}{2} \times 5\frac{1}{3}$:

$$2\frac{1}{2} \times 5\frac{1}{3} =$$ Change each mixed number to an improper fraction.

$$\frac{5}{2} \times \frac{16}{3} =$$ Reduce.

$$\frac{5}{\underset{1}{2}} \times \frac{\overset{8}{16}}{3} =$$

$$\frac{5}{1} \times \frac{8}{3} =$$ Multiply numerators. Multiply denominators.

$$\frac{40}{3} =$$ Change the product to a whole or mixed number.

$$13\frac{1}{3}$$

● **RELATED TOPICS: Improper fractions, Mixed numbers, Reducing fractions**

▲ **APPLICATION: Construction: Height of a Brick and Mortar Wall**

How to

Calculate the height of a brick and mortar construction:

1. Multiply the number of rows *(courses)* of brick by the thickness of 1 brick.
2. Multiply the number of courses of mortar by the thickness of each mortar joint.
3. Add the sum of the thicknesses of brick and mortar.

EXAMPLE   Bricks that are $2\frac{1}{4}$ in. thick form a brick wall with $\frac{3}{8}$-in. mortar joints. To find the height of the wall above the foundation after nine courses:

Unknown facts   Height of the wall after nine courses of brick have been laid.

Known facts $2\frac{1}{4}$ in.   Note the thickness of each brick.

$\frac{3}{8}$ in.   Note the thickness of each mortar joint.

9   Note the number of courses (or rows) of brick and mortar joints.

Relationships   Height of wall = Thickness of each mortar joint × Number of mortar joints + Thickness of each brick × Number of rows of brick

Estimation   Each brick is a little more than 2 in. thick, so the wall should be at least 2 × 9 or 18 in. high. Since the mortar joint is not quite $\frac{1}{2}$ in. and the fractional portion of the brick's thickness is less than $\frac{1}{2}$ in., the combined thickness of the brick and mortar joint must be less than 3 in. So the total height of the wall should be less than 3 × 9 or 27 in. We estimate the wall height to be between 18 and 27 in.

Calculation $\left(9 \times 2\frac{1}{4}\right) + \left(9 \times \frac{3}{8}\right)$

$\left(\frac{9}{1} \times \frac{9}{4}\right) + \left(\frac{9}{1} \times \frac{3}{8}\right)$

$\frac{81}{4} + \frac{27}{8}$

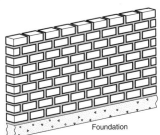

Foundation

$$20\frac{1}{4} + 3\frac{3}{8} \qquad \text{Find a common denominator.}$$

$$20\frac{2}{8} + 3\frac{3}{8} = 23\frac{5}{8} \quad \text{Add.}$$

Interpretation   The wall is $23\frac{5}{8}$ in. high.

⌞

● RELATED TOPIC: Six-step problem-solving plan

▲ APPLICATION: Increase or Decrease a Recipe

How to

Increase or decrease a recipe:

1. Determine the number of servings needed.
2. Determine the number of servings the original recipe yields.
3. Write the increase/decrease factor as

$$\frac{\text{number of servings needed}}{\text{number of servings yield in original recipe}}$$

4. Multiply each measure in the original recipe by the increase/decrease factor.

EXAMPLE   To make 4 dozen cookies using the recipe:

Easy Chocolate Cookies

    4 cups sugar
1 1/2 cups margarine or butter
1 1/3 cups evaporated milk
    2 small packages chocolate flavor
        instant pudding mix
    7 cups noninstant oatmeal
1 1/2 cups chocolate baking chips
    1 cup pecans or walnuts chopped

Bring sugar, margarine, and evaporated milk to a boil, stirring constantly. Boil for 1 minute, then remove from heat.

Stir in pudding mix, then oats, and mix well. Cool for 10 minutes, then stir in chocolate chips and nuts.

Drop by heaping teaspoonful onto waxed paper and let harden. Makes 8 dozen cookies.

4 dozen cookies are needed.
The recipe yields 8 dozen cookies.

$$\frac{4}{8} = \frac{1}{2}$$ Write decrease factor and multiply factor by the quantity of each ingredient.

$4 \times \dfrac{1}{2} = \dfrac{4}{1} \times \dfrac{1}{2} =$ $\quad 2 \quad$ cups sugar

$1\dfrac{1}{2} \times \dfrac{1}{2} = \dfrac{3}{2} \times \dfrac{1}{2} =$ $\quad \dfrac{3}{4}$ cup margarine

$1\dfrac{1}{3} \times \dfrac{1}{2} = \dfrac{4}{3} \times \dfrac{1}{2} =$ $\quad \dfrac{2}{3}$ cup evaporated milk

$2 \times \dfrac{1}{2} = \dfrac{2}{1} \times \dfrac{1}{2} =$ $\quad 1 \quad$ package chocolate instant pudding mix

$7 \times \dfrac{1}{2} = \dfrac{7}{1} \times \dfrac{1}{2} = \dfrac{7}{2} = 3\dfrac{1}{2}$ cups oatmeal

$1\dfrac{1}{2} \times \dfrac{1}{2} = \dfrac{3}{2} \times \dfrac{1}{2} =$ $\quad \dfrac{3}{4}$ cup chocolate chips

$1 \times \dfrac{1}{2} =$ $\quad \dfrac{1}{2}$ cup pecans or walnuts

Two numbers are **reciprocals** if their product is 1. The numbers $\frac{1}{2}$ and 2 are reciprocals. $\frac{2}{3}$ and $\frac{3}{2}$ are reciprocals. The **multiplicative inverse** of a number is its reciprocal. A number times its multiplicative inverse is 1, the **multiplicative identity.**

$$\frac{n}{1} \times \frac{1}{n} = 1$$

How to

Find the reciprocal of a number:

1. Write the number in fractional form.
2. Interchange the numerator and denominator so that the numerator is the denominator and the denominator is the numerator.

Interchanging the numerator and denominator of a fraction is commonly called **inverting** the fraction.

EXAMPLE   To find the reciprocal of $\frac{4}{7}$, $\frac{1}{5}$, 3, $2\frac{1}{2}$, 0.8, 1, and 0:

The reciprocal of $\frac{4}{7}$ is $\frac{7}{4}$ or $1\frac{3}{4}$.   Interchange the numerator and
                                             denominator.
The reciprocal of $\frac{1}{5}$ is $\frac{5}{1}$ or 5.   Write in fraction form. Interchange
                                             the numerator and denominator.
The reciprocal of 3 is $\frac{1}{3}$. $\left(3 = \frac{3}{1}\right)$ Write 3 as an improper fraction.

The reciprocal of $2\frac{1}{2}$ is $\frac{2}{5}$. $\left(2\frac{1}{2} = \frac{5}{2}\right)$ Write $2\frac{1}{2}$ as an improper fraction.

The reciprocal $\left(0.8 = \frac{8}{10} = \frac{4}{5}\right)$ is $\frac{5}{4}$ or **1.25.**   Write 0.8 as a common
                                                                         fraction.
The reciprocal of 1 is **1.** $\left(1 = \frac{1}{1}\right)$

0 has no reciprocal. $\left(0 = \frac{0}{1}\right) \frac{1}{0}$ **is undefined.**

● **RELATED TOPICS: Changing decimals to fractions,
Changing fractions to decimals, Decimals**

How to

Divide fractions:

1. Change the division to an equivalent multiplication by replacing the divisor with its reciprocal and replacing the division sign ($\div$) with a multiplication sign ($\times$).
2. Perform the resulting multiplication.

EXAMPLE   To find $\dfrac{5}{8} \div \dfrac{2}{3}$:

$$\frac{5}{8} \div \frac{2}{3} = \frac{5}{8} \times \frac{3}{2} = \frac{15}{16}$$

How to

Divide mixed numbers:

1. Change each mixed number or whole number to an improper fraction.
2. Convert to an equivalent multiplication problem using the reciprocal of the divisor.
3. Multiply according to the rule for multiplying fractions.

EXAMPLE   To find $2\frac{1}{2} \div 3\frac{1}{3}$:

$2\frac{1}{2} \div 3\frac{1}{3} =$ Change each mixed or whole number to an improper fraction.

$\frac{5}{2} \div \frac{10}{3} =$ Change division to equivalent multiplication.

$\overset{1}{\frac{5}{2}} \times \frac{3}{\underset{2}{10}} =$ Multiply.

$\frac{3}{4}$

EXAMPLE   To find $5\frac{3}{8} \div 3$:

$5\frac{3}{8} \div 3 =$ Change mixed number and whole number to an improper fraction.

$\frac{43}{8} \div \frac{3}{1} =$ Change division to equivalent multiplication.

$\frac{43}{8} \times \frac{1}{3} =$ Multiply.

$\frac{43}{24} = 1\frac{19}{24}$

● **RELATED TOPICS: Multiplying mixed numbers, Reciprocals**

▲ **APPLICATION: Subdividing Property**

EXAMPLE   A developer subdivides $5\frac{1}{4}$ acres into lots; each lot is $\frac{7}{10}$ of an acre. To determine how many lots are made:

$$5\frac{1}{4} \div \frac{7}{10} = \frac{21}{4} \div \frac{7}{10} = \frac{\overset{3}{21}}{\underset{2}{4}} \times \frac{\overset{5}{10}}{\underset{1}{7}} = \frac{15}{2} = 7\frac{1}{2}$$

Seven lots are made so that each is $\frac{7}{10}$ of an acre. The $\frac{1}{2}$ lot is left over or combined with one of the other lots.

▲ **APPLICATION: Accounting for Waste When Cutting Boards, Shelves, or Other Objects**

How to

Account for waste when cutting boards, shelves, or objects:

1. Determine the total number of equal pieces needed.
2. Determine the number of cuts required by subtracting 1 from the number of pieces needed.
3. Multiply the amount of waste for each cut by the number of cuts to find amount of total waste.
4. Subtract the amount of total waste from the length of the object.
5. Divide the result from Step 4 by the number of pieces needed to get the length of each piece.

EXAMPLE A piece of trophy column stock that is $21\frac{1}{2}$ in. long is cut into four equal trophy columns. If $\frac{1}{16}$ in. is wasted on each cut, the length of each piece can be found:

Use the Six-Step Problem-Solving Strategy.

Unknown facts   Length of cuts to be made.

Known facts   4   Note the number of pieces needed.

                     3   Note the number of cuts to be made $(4 - 1 = 3)$.

                     $\frac{1}{16}$ in.   Note the waste for each cut.

                     $21\frac{1}{2}$ in. Note the length of trophy column stock.

Relationships   Length of each piece = (Total length − 3 cuts × Amount wasted for each cut) ÷ 4 pieces of stock needed.

Estimation   If the stock is 20 in. long and cut into 4 pieces and if waste is disregarded, each piece will be 5 in.

Calculation Three cuts are to be made and each cut wastes $\frac{1}{16}$ in. Find the amount wasted.

$$\frac{1}{16} \times 3 = \frac{3}{16}$$ Total waste.

$$21\frac{1}{2} - \frac{3}{16}$$ Subtract to find the amount of stock that will be left to divide equally into four trophy columns.

$$21\frac{1}{2} = 21\frac{8}{16}$$ Align vertically and use common denominators.

$$-\ \frac{3}{16} = \frac{3}{16}$$ Amount of stock left to be divided.

$$21\frac{5}{16} \text{ in.}$$

Find the length of each trophy column.

$$21\frac{5}{16} \div 4 = \frac{341}{16} \div \frac{4}{1} = \frac{341}{16} \times \frac{1}{4} = \frac{341}{64} = 5\frac{21}{64}$$

Interpretation **Each trophy column is $5\frac{21}{64}$ in. long.**

⌞

● **RELATED TOPIC: Reading a rule**

A **complex fraction** is a fraction in which either the numerator or denominator or both contain a fraction or mixed number. A complex fraction is another way of writing division.

How to

Simplify a complex fraction:

1. Rewrite using the divided by symbol $\div$.
2. Perform the indicated division.

EXAMPLE To simplify $\dfrac{6\frac{3}{8}}{4\frac{1}{2}}$:

Fractions

$$\frac{6\frac{3}{8}}{4\frac{1}{2}} = \quad$$ Rewrite using the $\div$ symbol.

$$6\frac{3}{8} \div 4\frac{1}{2} = \quad$$ Perform the indicated division. Write mixed numbers as improper fractions.

$$\frac{51}{8} \div \frac{9}{2} = \quad$$ Change division to multiplication.

$$\frac{\overset{17}{\cancel{51}}}{\underset{4}{\cancel{8}}} \times \frac{\overset{1}{\cancel{2}}}{\underset{3}{\cancel{9}}} = \quad$$ Reduce and multiply.

$$\frac{17}{12} = 1\frac{5}{12} \quad$$ Write as a mixed number.

How to

Raise a fraction or quotient to a power:

1. Raise the numerator to the power.
2. Raise the denominator to the power.

$$\left(\frac{a}{b}\right)^n = \frac{a^n}{b^n} \qquad b \neq 0$$

EXAMPLE   Raise the fractions to the indicated power:

(a) $\left(\dfrac{2}{3}\right)^2$ (b) $\left(\dfrac{1}{2}\right)^3$

(a) $\left(\dfrac{2}{3}\right)^2 = \dfrac{2^2}{3^2} = \dfrac{4}{9}$ Raise the numerator to the power.
Raise the denominator to the power.

(b) $\left(\dfrac{1}{2}\right)^3 = \dfrac{1^3}{2^3} = \dfrac{1}{8}$ $(1)(1)(1) = 1;\ (2)(2)(2) = 8$

● **RELATED TOPIC: Laws of exponents**

CHAPTER 3

Decimals

- **3–1 Decimals and the Place-Value System**
- **3–2 Basic Operations with Decimals**

■ 3–1 Decimals and the Place-Value System

A decimal fraction is a fraction that has a denominator of 10 or some power of 10, such as 100 or 1000. Often the terms **decimal fraction, decimal number,** and **decimal** are used interchangeably.

The place on the right of the ones place is called the *tenths place*. A period (.), called the **decimal point,** is placed between the ones place and tenths place to distinguish between whole amounts and fractional amounts.

The digits to the right of the ones place represent the numerator of the fraction. The place value of the rightmost digit indicates the denominator.

How to

Identify the place value of digits in decimal fractions:

1. Mentally position the decimal number on the decimal place-value chart so that the decimal point of the number aligns with the decimal point on the chart.
2. Identify the place value of each digit according to its position on the chart.

EXAMPLE   To identify the place value of each digit in 32.4675:
Position the number 32.4675 on a place-value chart.
3 is in the tens place.
2 is in the ones place.
4 is in the tenths place.
6 is in the hundredths place.
7 is in the thousandths place.
5 is in the ten-thousandths place.

| Billions (1,000,000,000) |
| Hundred millions (100,000,000) |
| Ten millions (10,000,000) |
| Millions (1,000,000) |
| Hundred thousands (100,000) |
| Ten thousands (10,000) |
| Thousands (1,000) |
| Hundreds (100) |
| **3** Tens (10) |
| **2** Ones (1) |
| **.** |
| **4** Tenths ($\frac{1}{10}$ or 0.1) |
| **6** Hundredths ($\frac{1}{100}$ or 0.01) |
| **7** Thousandths ($\frac{1}{1,000}$ or 0.001) |
| **5** Ten-thousandths ($\frac{1}{10,000}$ or 0.0001) |
| Hundred-thousandths ($\frac{1}{100,000}$ or 0.00001) |
| Millionths ($\frac{1}{1,000,000}$ or 0.000001) |
| Ten-millionths ($\frac{1}{10,000,000}$ or 0.0000001) |
| Hundred-millionths ($\frac{1}{100,000,000}$ or 0.00000001) |
| Billionths ($\frac{1}{1,000,000,000}$ or 0.000000001) |

• Decimal point

● RELATED TOPIC: Whole number place-value chart

Decimals

How to

Read and write a decimal number in words:

1. Mentally align the number on the decimal place-value chart so that the decimal point of the number is directly under the decimal point on the chart.
2. Read the whole-number part.
3. Use *and* for the decimal point only if there is a whole-number part.
4. Read the decimal part the same way you read a whole number.
5. End by reading the *place value* of the rightmost digit in the decimal part.

EXAMPLE   To read 52.386 and write it in words:

1. Mentally align the number with the chart.
2. Read the whole-number part.

52. 386

fifty-two

3. Use *and* to separate the whole-number part from the decimal part.

52.386

and

4. Read the decimal part as you would read a whole number.

52.386

three hundred eighty-six

5. End by reading the *place value* of the rightmost digit in the decimal part.

52.386

thousandths

52.386 is "fifty-two and three hundred eighty-six thousandths."

● **RELATED TOPIC: Read and write whole numbers**

Informal use of the word point

Informally, the decimal point is sometimes read as "point." Thus, 3.6 is read "three point six." The decimal 0.0162 can be read as "point zero one six two." This informal process is often used in communication to ensure that numbers are not miscommunicated.

Unwritten decimals

When we write whole numbers using numerals we usually omit the decimal point; the decimal point is understood to be at the end of the whole number. Therefore, any whole number, such as 32, can be written without a decimal (32) or with a decimal (32.).

How to

Write a fraction with a power-of-10 denominator as a decimal number:

1. Observe the number of zeros in the denominator to find the number of decimal places.

10	→ 1 place
100	→ 2 places
1,000	→ 3 places
10,000	→ 4 places

   etc.

   To write $\frac{17}{1,000}$ as a decimal:

   0._ _ _

2. Place the numerator so that the last digit is in the farthest place on the right.

   0._ 1 7

3. Fill in any blank spaces with zeros.

   0.017

Do ending zeros change the value of a decimal number?

When we attach zeros on the *right* end of a decimal number, we do not change the value of the number.

$$0.5 = 0.50 = 0.500 \qquad \frac{5}{10} = \frac{50}{100} = \frac{500}{1,000}$$

How to

Change a decimal number to a fraction or mixed number in lowest terms:

1. Write the numerator as the digits without the decimal point.
2. Write the denominator as a power of 10 with as many zeros as there are places after the decimal point.
3. Reduce and, if the fraction is improper, convert to a mixed number.

EXAMPLE   To write 0.4 and 0.075 as fractions:

$$0.4 = \frac{4}{10} = \frac{2}{5}$$ Tenths indicates a denominator of 10.

$$0.075 = \frac{75}{1,000} = \frac{3}{40}$$ Thousandths indicates a denominator of 1,000.

⌐

● **RELATED TOPIC: Reducing fractions**

How to

Change a fraction to a decimal number:

1. Place a decimal point after the numerator.
2. Attach zeros to the numerator as needed for division.
3. Divide the numerator by the denominator using long division.

Decimals

EXAMPLE To change $\frac{7}{8}$ to a decimal:

$$7 \div 8 \quad \text{or} \quad 8\overline{)7.000}$$

$$\begin{array}{r} 0.875 \\ \underline{6\ 4} \\ 60 \\ \underline{56} \\ 40 \\ \underline{40} \end{array}$$

Attach zeros until the division terminates; that is, it has no remainder.

How to

Compare decimal numbers:

1. Compare whole-number parts.
2. If the whole-number parts are equal, compare digits place by place, starting at the tenths place and moving to the right.
3. Stop when the digits in the same place for the two numbers are different.
4. The digit that is larger determines the larger decimal number.

EXAMPLE To compare the two numbers to see which is larger:

32.47   32.48

1. Look at the whole-number parts: They are the same.
2. Look at the tenths place for each number. Both numbers have a 4 in the tenths place.
3. Look at the hundredths place. The digits are different and 8 is larger than 7.
4. **32.48 is the larger number.**

EXAMPLE To write two inequalities for the numbers 0.4 and 0.07:
Since the whole-number parts are the same (0), we compare the digits in the tenths place. 0.4 is larger because 4 is larger than 0.

$0.4 > 0.07$ or $0.07 < 0.4$

Common denominators in decimals

Decimal fractions have a common denominator if they have the same number of digits to the right of the decimal point. To compare 0.4 and 0.07 using common denominators, write 0.4 as 0.40. Then, 40 hundredths is more than 7 hundredths.

● **RELATED TOPICS:** Common denominators, Fractions, Inequalities, Whole numbers

How to

Round a decimal number to a given place value:

1. Locate the digit that occupies the rounding place. Examine the digit to the immediate right.
2. If the digit to the right of the rounding place is 0, 1, 2, 3, or 4, do not change the digit in the rounding place. If the digit to the right of the rounding place is 5, 6, 7, 8, or 9, add 1 to the digit in the rounding place.
3. Replace all digits to the right of the rounding place with zeros up to the decimal point. Drop any digits that are to the right of the digit in the rounding place and that follow the decimal point.

Decimals

EXAMPLE   To round 46.879 to the hundredths place:

46.8 7 9   7 is in the hundredths place.

46.8 7 9   The next digit to the right is 9, so add 1 to 7 (7 + 1 = 8).

46.8 8   Write all digits to the left of 7 as they are. Drop the 9 because it is to the right of the rounding place and to the right of the decimal.

46.88

Rounding to dollars or cents

When we round to the nearest dollar, we are rounding to the *ones* place.

$293.48 rounds to $293.

One cent is 1 hundredth of a dollar, so to round to the nearest cent is to round to the *hundredths* place.

$71.8986 rounds to $71.90.

> **How to**
>
> *Round a decimal number to a number with one nonzero digit:*
>
> 1. Find the first nonzero digit from the left.
> 2. Round the number to the place value of the first nonzero digit.
> 3. Replace all digits to the right of the rounding place with zeros up to the decimal point. Drop any digits that are to the right of the digit in the rounding place *and* that follow the decimal point.

EXAMPLE   To round 78.4 to one nonzero digit:

78.4   The first nonzero digit is 7.

7 8.4

7 8.4   Add 1 to 7 because the next digit to the right is 5 or more.

8

80   Replace the digit to the right of 8 with a 0. Drop the digits after the decimal point.

● **RELATED TOPICS: Estimation, Rounding whole numbers**

■ 3–2 Basic Operations with Decimals

> **How to**
>
> *Add decimals:*
>
> 1. Arrange the numbers so that the decimal points are in one vertical column.
> 2. Add each column.
> 3. Align the decimal for the sum in the same vertical column.

EXAMPLE   To add 42.3 + 17 + 0.36:

42.3
17   Note that the decimal in 17 is understood to be at the right end.
 0.36

We should be very careful in aligning digits. Careless writing can cause unnecessary errors. To avoid difficulty in adding the columns, we may write each number so that all have the same number of decimal places by attaching zeros on the right.

42.3 0
17. 00
 0.36
59.66
└─

How to

Subtract decimals:

1. Arrange the numbers so that the decimal points align vertically.
2. Subtract each column beginning at the right.
3. Interpret blank places as zeros.
4. Place the decimal point in the difference with the same vertical alignment.

EXAMPLE   To subtract 8.29 from 13.76:

13.76   Notice how the decimal points are aligned in a vertical
− 8.29   column.
 5.47
└─

EXAMPLE   To subtract 7.18 from 15:

15.   Because 15 is a whole number, the decimal follows the 5.
− 7.18

15. 00   Interpret blanks as zeros. Borrow (regroup).
− 7.18
 7.82
└─

● **RELATED TOPICS: Adding whole numbers, Subtracting whole numbers**

How to

Multiply decimals:

1. Align the numbers as if they were whole numbers and multiply.
2. Count the total number of digits to the right of the decimal in both factors.
3. Place the decimal in the product so that the number of decimal places is the sum of the number of decimal places in the factors.

EXAMPLE To multiply 1.36 × 0.2:

$$
\begin{array}{r}
1.36 \\
\times \quad 0.2 \\
\hline
0.272
\end{array}
$$

Note that in multiplication the decimals do *not* have to be in a straight line. Place a zero in the ones place so that the decimal point will not be overlooked.

EXAMPLE To multiply 0.309 × 0.17:

$$
\begin{array}{r}
0.309 \\
\times \quad 0.17 \\
\hline
2163 \\
309 \quad \\
\hline
0.0\,5253
\end{array}
$$

We needed 5 decimal digits but had only 4, so we inserted a zero on the *left* to give the appropriate number of decimal places.

● **RELATED TOPIC: Multiplying whole numbers**

How to

Divide decimals:

1. Move the decimal in the divisor if necessary so that it is on the right side of all digits. (By moving the decimal, you are multiplying by 10, 100, 1000, and so on.)
2. Move the decimal in the dividend to the right as many places as the decimal was moved in the divisor. Attach zeros if necessary. (This is multiplying the dividend by the same number as was used in step 1.)
3. Write the decimal point in the answer directly above the new position of the decimal in the dividend. (Do this *before* dividing.)
4. Divide as you would in whole numbers.

EXAMPLE To divide 4.8 ÷ 6:

$$
6\overline{)4.8}
$$

Insert a decimal point above the decimal point in the dividend.

When the divisor is a whole number, the decimal is understood to be to the right of 6 and is not moved. The decimal is placed in the quotient directly above the decimal in the dividend.

$$\begin{array}{r} 0.8 \\ 6\overline{)4.8} \end{array}$$ Divide.

EXAMPLE   To divide 3.12 ÷ 1.2:

$$1.2\overline{)3.1\,2}$$ Move the decimal in the divisor and dividend.

$$12\overline{)31.2}$$ Write the decimal in the quotient, then divide.

$$\begin{array}{r} 2.6 \\ 12\overline{)31.2} \\ \underline{24} \\ 7\,2 \\ \underline{7\,2} \end{array}$$

How to

Round a quotient to a place value:

1. Divide to one place past the desired rounding place.
2. Attach zeros to the dividend after the decimal if necessary to carry out the division.
3. Round the quotient to the place specified.

EXAMPLE   To divide and round the quotient to the nearest tenth:

$$3.2\overline{)15.27}$$

$$\begin{array}{r} 4.77 \\ 3.2\overline{)15.2\,70} \end{array}$$ Since you are rounding to the nearest tenth, divide to the hundredths place.

$$\begin{array}{r} 12\,8 \\ \hline 2\,4\,7 \\ 2\,2\,4 \\ \hline 2\,30 \\ 2\,24 \\ \hline 6 \end{array}$$

4.77 rounds to 4.8.

● **RELATED TOPICS:** Average, Dividing whole numbers, Mean

> **How to**
>
> *Multiply a number by a power of 10:*
>
> 1. Move the decimal point in the number to the *right* as many places as the 10, 100, 1000, or so on, has zeros.
> 2. Attach zeros on the right if necessary.

EXAMPLE   To multiply 237 × 100:

$237.00 \times 100 = \mathbf{23{,}700}$ Attach two zeros to the *right* of the 7 in 237 and insert the appropriate comma.

L

EXAMPLE   To multiply 36.2 × 1,000:

$36.200 \times 1000 = \mathbf{36{,}200}$ Move the decimal point three places to the *right*. Two zeros need to be attached.

L

> **How to**
>
> *Divide a number by a power of 10:*
>
> 1. Move the decimal to the *left* as many places as the divisor has zeros.
> 2. Attach zeros to the left if necessary.
> 3. You may drop zeros to the right of the decimal point if they follow the last nonzero digit of the quotient.

EXAMPLE   To divide 63 by 100:

$63 \div 100 = 0.63 = 0.63$ Decimal is after 3 in 63. Move decimal two places to the left.

L

● **RELATED TOPIC:** Multiply by factors with ending zeros

CHAPTER 4

Integers and Signed Numbers

- ■ **4–1 Natural Numbers, Whole Numbers, and Integers**
- ■ **4–2 Basic Operations with Integers**
- ■ **4–3 Signed Fractions and Decimals**

■ 4–1 Natural Numbers, Whole Numbers, and Integers

The set of counting numbers, or **natural numbers,** begins with the numbers 1, 2, 3, 4, 5, 6, 7, and continues indefinitely. The natural numbers and zero form a set of numbers called **whole numbers.**

When the opposite of each natural number is included, the set of whole numbers can be expanded to form the set of **integers.** A number line for integers continues indefinitely in both directions.

The **opposite** of a number is the same number of units from zero, but in the opposite direction. The opposite of 3 is -3. The opposite of -5 is 5. Zero has no opposite.

The positive sign for positive values may be included to draw attention to the sign or to emphasize its direction, but negative signs must always be included.

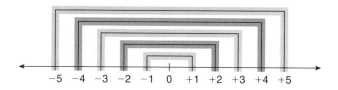

How to

Compare two integers:

1. Visualize the integers on a number line.
2. The leftmost integer is the smaller integer. The rightmost integer is the larger integer.
3. Use the appropriate symbol, $<$ (less than) or $>$ (greater than), to write an inequality.

EXAMPLE   The symbols $<$ and $>$ can be used to indicate whether the first number is less than or greater than the second number. (a) 7 ___ 9 (b) 0 ___ -1 (c) -4 ___ -2

(a) $7 < 9$ 7 is smaller—to the left of 9 on the number line.

(b) $0 > -1$ 0 is larger—to the right of -1 on the number line.

(c) $-4 < -2$ -4 is smaller—to the left of -2 on the number line.

L_

● **RELATED TOPIC: Compare whole numbers**

The **absolute value** of a number is described as the number of units of distance the number is from zero. The symbol for absolute value is | |; |3| is read "the absolute value of 3."

Distance is a physical property and cannot have a negative value. The absolute value of a number is always **nonnegative** (zero or positive).

How to

Find the absolute value of a number:

1. If the number is positive:
   The absolute value of a positive number is the number.

   $|a| = a$,   for $a > 0$ $|5| = 5$

2. If the number is negative:
   The absolute value of a negative number is equal to its opposite.

   $|a| = -a$,   for $a < 0$ $|-3| = 3$ (3 is the opposite of -3.)

3. If the number is zero:
   The absolute value of zero is zero.

   $|0| = 0$

EXAMPLE   To find the absolute values of the quantities (a) $|9|$ (b) $|-4|$ (c) $|0|$:

(a) $|9| = 9$ 9 is positive. Its absolute value is the number itself.

(b) $|-4| = 4$ −4 is negative. Its absolute value is the opposite of −4.

(c) $|0| = 0$ 0 is unsigned. Its absolute value is still zero.

The **opposite** of a number is also called the **additive inverse** of the number. A number and its additive inverse (opposite) have the same absolute value but different directional signs.

How to

Find the opposite of a number:

1. If the number is positive, its opposite is the negative number with the same absolute value.
2. If the number is negative, its opposite is the positive number with the same absolute value.

EXAMPLE   To find the opposite of (a) −8 (b) 6:

(a)   The opposite of −8 is 8. $|-8|$ is 8.

(b)   The opposite of 6 is −6. $|6|$ is 6.

■ 4–2 Basic Operations with Integers

How to

Add integers (signed numbers) with like signs:

1. Add the absolute values of the numbers.
2. Give the sum the common or like sign.

EXAMPLE   To add (a) $3 + 6$ (b) $-5 + (-8)$:

(a) $3 + 6$ Add absolute values. Keep common positive sign.

      +9 or **9**

(sidebar) Signed Numbers

(b) $-5 + (-8)$ Add absolute values (5 + 8 = 13). Keep common
negative sign.
$$-13$$

⌐

Addition is a **binary operation.** This means that only two numbers are used in the operation. If more than two numbers are involved, two are added and then the next number is added to the sum of the first two.

EXAMPLE To add $-3 + (-4) + (-1)$:

$-3 + (-4)$ $+ (-1) =$ Add the first two numbers.

-7 $+ (-1) = -8$ Add the sum to the remaining number.

⌐

How to

Add integers (signed numbers) with unlike signs:

1. *Subtract* the smaller absolute value from the larger absolute value.
2. Give the sum the sign of the number with the larger absolute value.

EXAMPLE To add $7 + (-12)$:

$7 + (-12) =$ Signs are unlike.

$7 + (-12) = -5$ Subtract absolute values: 12 − 7 = 5. Keep the sign
of the −12 (negative), the larger absolute value.

⌐

TIP

To add, you . . . subtract?

When adding integers, phrases used in arithmetic, such as "find the sum" or "add," may be misleading. Yes, in the operation of addition, sometimes we add absolute values (when signs are alike) and sometimes we subtract absolute values (when signs are different). The word "combine" can be used to imply the addition of numbers with either like or unlike signs.

EXAMPLE   To combine $-5 + 7$:

$-5 + 7 =$ Signs are unlike.

$-5 + 7 = 2$ Subtract absolute values: $7 - 5 = 2$. Keep the sign of the
⌐   7 (positive), the larger absolute value.

Zero is the **additive identity** for all numbers.

$a + 0 = 0 + a = a$ for all values of a.

$-5 + 0 = -5$

The opposite of a number is the **additive inverse** of the number. A number added to its additive inverse equals the additive identity.

$a + (-a) = (-a) + a = 0$ for all values of a.

$7 + (-7) = 0$

● **RELATED TOPICS: Additive identity for whole numbers, Commutative property for addition**

How to

Subtract integers (signed numbers):

1. Add the opposite of the subtrahend (second number) to the minuend (first number).
2. Apply the appropriate rule for adding signed numbers with like or unlike signs.

EXAMPLE   To subtract (a) $2 - 6$ (b) $-9 - 5$ (c) $12 - (-4)$
(d) $-7 - (-9)$:

(a) $2 - 6 =$ Subtract positive 6 from positive 2. Write sign of
                            subtrahend.

       $2 - (+6) =$ Add the opposite of 6, which is -6, to 2.

       $2 + (-6) =$ Apply the rule for adding numbers with unlike signs.

       $2 + (-6) = -4$

(b) $-9 - 5 =$ Subtract positive 5 from negative 9. Write sign of
                            subtrahend.

       $-9 - (+5) =$ Add the opposite of 5, which is -5, to -9.

       $-9 + (-5) =$ Apply the rule for adding numbers with like signs.

       $-9 + (-5) = -14$

**Signed
Numbers**

(c) $12 - (-4) =$ Subtract negative 4 from positive 12. Add the opposite of -4, which is $+4$, to 12.

    $12 + (+4) =$ Apply the rule for adding numbers with like signs.

       $12 + 4 = 16$

(d) $-7 - (-9) =$ Subtract negative 9 from negative 7. Add the opposite of -9, which is $+9$, to -7.

    $-7 + (+9) =$ Apply the rule for adding numbers with unlike signs.

       $-7 + 9 = 2$

Writing subtractions as equivalent additions

When writing a subtraction as an equivalent addition, you make *two* changes.

1. Change the operation from subtraction to addition.
2. Change the subtrahend to its opposite.

$2 - (+6)$	$-9 - (+5)$	$12 - (-4)$	$-7 - (-9)$
$2 + (-6)$	$-9 + (-5)$	$12 + (+4)$	$-7 + (+9)$
↑ ↑	↑ ↑	↑ ↑	↑ ↑
1. 2.	1. 2.	1. 2.	1. 2.

Omitting signs

Writing mathematical expressions that include every operational and directional sign is cumbersome. In general practice, we omit as many signs as possible. When two signs are written between two numbers, we can write a simplified expression with only one sign.

Plus, Plus: $+3 + (+5)$ Add $+3$ and $+5$. $3 + 5 = 8$

Minus, Minus: $+3 - (-5)$ Change to addition.
                            Add $+3$ and $+5$. $3 + 5 = 8$

Plus, Minus: $+3 + (-5)$ Add $+3$ and -5. $3 - 5 = -2$

Minus, Plus: $+3 - (+5)$ Change to addition.
                            Add $+3$ and -5. $3 - 5 = -2$

How to

Add and subtract more than two numbers:

1. Rewrite the problem so that all integers are separated by only one sign.
2. Add the signed numbers from left to right.

EXAMPLE   To evaluate (a) $3 - (-5) - 6$ (b) $-8 + 10 - (-7)$:

(a) $3 - (-5) - 6 =$ Rewrite with only one sign between integers.

   $3 + 5 - 6 =$ Add first two addends.

   $8 - 6 = 2$ Apply the rule for adding numbers with unlike signs.

(b) $-8 + 10 - (-7) =$ Rewrite with only one sign between integers.

   $-8 + 10 + 7 =$ Add first two addends.

   $2 + 7 = 9$ Apply the rule for adding numbers with like signs.

⌐

▲ APPLICATION: Net Profit or Net Loss

How to

Find the net profit or loss:

1. Interpret profits as positive and losses as negative.
2. Combine the signed numbers using the appropriate rule for adding signed numbers.
3. Interpret the result as a profit or loss.

EXAMPLE   A landscaping business makes a profit of $345 one week, has a loss of $34 the next week, and makes a profit of $235 the third week. To find the net profit:
   The *net* profit is the sum of the profits and losses.

Signed
Numbers

$345 - $34 + $235   Interpret profits as positive and losses as
profit   loss   profit   negative.

345 + 235 - 34   Rearrange and add positives.

580 - 34 = 546   Apply rule for adding numbers with unlike
signs.

The net profit for the three days was $546. Interpret answer.

How to

Multiply two integers (signed numbers):

1. Multiply the absolute values of the numbers as in whole numbers.
2. If the factors have like signs, the sign of the product is positive.
3. If the factors have unlike signs, the sign of the product is negative.

EXAMPLE To multiply (a) $-12(-2)$ (b) $10 \cdot 3$ (c) $25(-3)$
(d) $-5 * 7$:

(a) $-12(-2) = \mathbf{24}$ Like signs give a positive product.

(b) $10 \cdot 3 = \mathbf{30}$ Like signs give a positive product.

(c) $25(-3) = \mathbf{-75}$ Unlike signs give a negative product.

(d) $-5 * 7 = \mathbf{-35}$ Unlike signs give a negative product.

The **multiplicative identity** of a number is 1.

$a \cdot 1 = 1 \cdot a = a$ for all values of a.

Multiplication, like addition, is a **binary operation,** so we multiply two factors at a time.

EXAMPLE To multiply $4(-2)(6)$:

$4(-2)(6)$ Multiply the first two factors and apply the rule for factors with unlike signs.

$-8(6) = \mathbf{-48}$ Multiply and apply the rule for factors with unlike signs.

How to

Determine the sign of an integer raised to a natural-number power:

1. A positive number raised to any natural-number power is positive.
2. A negative number raised to an *even* natural-number power is positive.
3. A negative number raised to an *odd* natural-number power is negative.

EXAMPLE   To evaluate the powers (a) 4^3 (b) $(-2)^4$ (c) $(-3)^5$:

(a) $4^3 = 4(4)(4) = \mathbf{64}$

(b) $(-2)^4 = (-2)(-2)(-2)(-2) = \mathbf{16}$

(c) $(-3)^5 = (-3)(-3)(-3)(-3)(-3) = \mathbf{-243}$

 TIP

Negative base versus an opposite

$(-2)^4$ is not the same expression as -2^4.

$$(-2)^4 = (-2)(-2)(-2)(-2) = 16 \qquad -2^4 = -(2)(2)(2)(2) = -16$$

$(-2)^4$ is a negative base raised to a power. -2^4 is the opposite of 2^4.
   Sometimes the values of two expressions are equal, but the interpretation is different. The expression $(-3)^5$ means "negative three raised to the fifth power." The expression -3^5 is a different expression but has the same value because the exponent is an odd number.

$$(-3)^5 = (-3)(-3)(-3)(-3)(-3) = -243$$
$$-3^5 = -(3)(3)(3)(3)(3) = -243$$

How to

Evaluate powers that have zero as a base or exponent:

1. Zero raised to any natural-number power is zero.

$0^n = 0; n \neq 0$

2. Any nonzero number raised to the zero power is 1.

$n^0 = 1; n \neq 0$

EXAMPLE   To evaluate the powers (a) 0^8 (b) $(-7)^0$:

(a) $0^8 = 0$ Zero raised to any natural-number power is zero.

(b) $(-7)^0 = 1$ A nonzero number raised to the zero power is 1.

⌊

How to

Divide two integers (signed numbers):

1. Divide the absolute values of the numbers as in whole numbers.
2. If the values have like signs, the sign of the quotient is positive.
3. If the values have unlike signs, the sign of the quotient is negative.

EXAMPLE   To divide (a) $\dfrac{-8}{-2}$ (b) $\dfrac{6}{3}$ (c) $\dfrac{10}{-2}$ (d) $\dfrac{-9}{1}$:

(a) $\dfrac{-8}{-2} = 4$ Like signs give a positive quotient.

(b) $\dfrac{6}{3} = 2$ Like signs give a positive quotient.

(c) $\dfrac{10}{-2} = -5$ Unlike signs give a negative quotient.

(d) $\dfrac{-9}{1} = -9$ Unlike signs give a negative quotient.

⌊

■ 4–3 Signed Fractions and Decimals

How to

Find an equivalent signed fraction:

1. Identify the three signs of the fraction (sign of the fraction, sign of the numerator, and sign of the denominator).
2. Change any two of the three signs to the opposite sign.

EXAMPLE   To change $-\dfrac{-2}{-3}$ to three equivalent signed fractions:

$$-\frac{-2}{-3} = +\frac{+2}{-3}$$ Change the sign of the fraction and the sign of the numerator.

$$-\frac{-2}{-3} = +\frac{-2}{+3}$$ Change the signs of the fraction and the denominator.

$$-\frac{-2}{-3} = -\frac{+2}{+3}$$ Change the signs of the numerator and the denominator.

Why would we want to change the signs of a fraction?

When changing any two signs of a fraction, we can accomplish these desirable outcomes.

- Avoid dealing with negatives.

$$-\frac{-3}{+4} = +\frac{+3}{+4}$$

$$+\frac{-6}{-7} = +\frac{+6}{+7}$$

$$-\frac{+5}{-9} = +\frac{+5}{+9}$$

- Change subtraction to addition.

$$-\frac{+5}{+6} = +\frac{-5}{+6}$$

- Avoid negative denominators.

$$-\frac{+3}{-4} = +\frac{+3}{+4}$$

$$+\frac{+2}{-5} = +\frac{-2}{+5}$$

- Deal with fewer negatives.

$$-\frac{-5}{-8} = +\frac{-5}{+8}$$

Signed Numbers

How to

Add and subtract signed fractions:

1. Change the signs of the fractions so the denominators are positive and subtractions are expressed as addition.
2. Apply the appropriate rules for adding integers (signed numbers) and for adding fractions.

EXAMPLE   To add $\dfrac{-3}{4} + \dfrac{5}{-8}$:

$\dfrac{-3}{4} + \dfrac{-5}{8}$ Change the signs of the numerator and denominator in the second fraction so that both denominators are positive.

$\dfrac{-6}{8} + \dfrac{-5}{8}$ Change to equivalent fractions with a common denominator.

$\dfrac{-11}{8}$ Add numerators, applying the rule for adding numbers with like signs.

$-1\dfrac{3}{8}$ Change to mixed number. The sign of the mixed number is determined by the rule for dividing numbers with unlike signs.

⌞

● **RELATED TOPICS: Adding fractions, Adding integers**

EXAMPLE   To subtract $\dfrac{-3}{7} - \dfrac{-5}{7}$:

$\dfrac{-3}{7} + \dfrac{5}{7}$ Change subtraction to addition by changing the signs of the second fraction by changing the signs of the fraction and the numerator.

$\dfrac{2}{7}$ Apply the rule for adding numbers with unlike signs.

⌞

How to

Multiply or divide signed fractions:

1. Write divisions as equivalent multiplications.
2. Apply the appropriate rules for multiplying integers (signed numbers) and for multiplying fractions.

EXAMPLE   To multiply $\left(\dfrac{-4}{5}\right)\left(\dfrac{3}{-7}\right)$:

$\dfrac{-4}{5} \cdot \dfrac{3}{-7}$ Multiply numerators and denominators.

$\dfrac{-12}{-35}$ Apply rule for dividing numbers with like signs.

$\dfrac{12}{35}$

⌞

How to

Perform basic operations with signed decimals:

1. Determine the indicated operations.
2. Apply the appropriate rules for signed numbers and for decimals.

EXAMPLE   To add -5.32 and -3.24:

$\begin{array}{r} -5.32 \\ -3.24 \\ \hline -8.56 \end{array}$ Align decimals and use the rule for adding numbers with like signs.

EXAMPLE   To subtract -3.7 from 8.5:

$8.5 - (-3.7) = 8.5 + 3.7 = \mathbf{12.2}$

Integers and signed numbers follow the same order of operations as whole number, fraction, and decimal computations.

● **RELATED TOPIC: Order of operations for whole numbers**

EXAMPLE   To evaluate $4 + 5(2 - 8)$:

$4 + 5\ (2 - 8)$ Perform operations in parentheses.

$4 + 5\ (-6)$ Multiply.

$4 - 30$ Add integers with unlike signs.

$4 - 30 = \mathbf{-26}$

EXAMPLE   To evaluate $5 + (-2)^3 - 3(5)$:

$5 + (-2)^3 - 3(5)$ Raise to a power.

$5 + (-8) - 3(5)$ Multiply.

$5 + (-8) - 15$ Change to one sign only between numbers.

$5 - 8 - 15$ First addition of integers.

$-3 - 15 =$ Remaining addition of integers.

$\mathbf{-18}$

Signed Numbers

▲ **APPLICATION: Stock Prices**

How to

Find the stock price of the previous day:

1. Determine the current day's closing price from a listing (often labeled *closing* or *last* price).
2. Determine the current day's change in price from the previous day (often labeled *change* or *chg*).
3. Subtract the amount of change from the current day's closing price.

   Previous day's price = Current day's price − change

EXAMPLE   ACE Ltd stock closed at $38.06, which was down $0.94 (−$0.94) from the previous day. What was the previous day's closing price?

$38.06 − (− $0.94) =   Current price − change

$38.06 + (+ $0.94) =   Change subtraction to addition.

$39.00   Previous day's price.

CHAPTER 5

Irrational Numbers and Real Numbers

- ■ **5–1 Root Notation**
- ■ **5–2 Irrational Numbers**

■ 5–1 Root Notation

The opposite or inverse operation for raising to powers is finding roots. A **square root** is the number used as a factor 2 times to give a certain number. The **radical notation** for square root is $\sqrt{}$, $\sqrt{25} = 5$. The **index** of a square root is 2, but is not required in radical notation.

A **cube root** is the number used as a factor 3 times to give a certain number. The index of a cube root is 3. For cube roots, the index 3 is written in the $\sqrt{}$ portion of the radical sign: $\sqrt[3]{8} = 2$. For **fourth roots**, the index 4 is written in the $\sqrt{}$ portion of the radical sign: $\sqrt[4]{81} = 3$.

Numbers that are squared, cubed, raised to the fourth power, and so on are called **perfect powers**. The principal (or positive) root of a perfect power is a whole number.

Powers of 2	Powers of 3	Powers of 4	Powers of 5
$2^1 = 2$	$3^1 = 3$	$4^1 = 4$	$5^1 = 5$
$2^2 = 4$	$3^2 = 9$	$4^2 = 16$	$5^2 = 25$
$2^3 = 8$	$3^3 = 27$	$4^3 = 64$	$5^3 = 125$
$2^4 = 16$	$3^4 = 81$	$4^4 = 256$	$5^4 = 625$
$2^5 = 32$	$3^5 = 243$	$4^5 = 1,024$	$5^5 = 3,125$
$2^6 = 64$	$3^6 = 729$	$4^6 = 4,096$	$5^6 = 15,625$

Selected roots

$\sqrt[4]{16} = 2$	$\sqrt[5]{243} = 3$	$\sqrt[3]{64} = 4$	$\sqrt[6]{15,625} = 5$

Another notation for indicating roots is a fractional exponent. To indicate a square root, use the exponent $\frac{1}{2}$. To indicate a cube root, use the exponent $\frac{1}{3}$. For a fourth root, use the exponent $\frac{1}{4}$. To generalize, the exponential notation for the root of a number is the number with the fractional exponent. The index of the root is the denominator of the fractional exponent.

$\sqrt[n]{a} = a^{\frac{1}{n}}$ where n is a natural number greater than 1 and a is a positive number.

How to

Write roots using radical and exponential notation:

1. Identify the index of the root.
2. For radical notation, place the index of the root in the $\sqrt{}$ portion of the radical sign with the radicand under the bar portion of the sign.
3. For exponential notation, write the radicand as the base and the index of the root as the denominator of a fractional exponent that has a numerator is 1.

TIP

Equivalent notations for roots

Just as $\frac{1}{2}$, 0.5, and 50% are three notations for writing equivalent values, radical notation and exponential notation are notations for writing equivalent values for roots.

In Words	Radical Notation	Exponential Notation
Square root of n	$\sqrt{n}$	$n^{1/2}$ or $n^{\frac{1}{2}}$ or $n^{0.5}$
Cube root of n	$\sqrt[3]{n}$	$n^{1/3}$ or $n^{\frac{1}{3}}$ or $n^{0.\overline{3}}$
Fourth root of n	$\sqrt[4]{n}$	$n^{1/4}$ or $n^{\frac{1}{4}}$ or $n^{0.25}$

EXAMPLE   To write the roots using radical notation and exponential notation:

   (a)   Square root of 25   (b)   Cube root of 125
   (c)   Fourth root of 625

   (a)   Square root of 25 $= \sqrt{25} = 25^{1/2}$ or $25^{0.5}$

(b)   Cube root of $125 = \sqrt[3]{125} = 125^{1/3}$ or $125^{0.\overline{3}}$

(c)   Fourth root of $625 = \sqrt[4]{625} = 625^{1/4}$ or $625^{0.25}$

■ 5–2 Irrational Numbers

Not all numbers are perfect powers. The root of a nonperfect power is an **irrational number.**

> ### How to
>
> #### Find the two whole numbers that are closest to the value of an irrational number:
>
> 1. Make a list of perfect powers that goes beyond the given number.
> 2. Identify the two perfect powers that the given number is between.
> 3. Find the principal roots of the two perfect powers from Step 2.
> 4. The root of the given number is between the roots found in Step 3.

EXAMPLE   To find two whole numbers that are closest to (a) $\sqrt{75}$ and (b) $\sqrt[3]{120}$:

(a)   1, 4, 9, 16, 25, 36, 49, 64, 81   Perfect squares. 75 is between 64 and 81.

$\sqrt{64} = 8$. $\sqrt{81} = 9$.

$\sqrt{75}$ **is between 8 and 9.**

(b)   1, 8, 27, 64, 125   Perfect cubes. 120 is between 64 and 125.

$\sqrt[3]{64} = 4$. $\sqrt[3]{125} = 5$.

$\sqrt[3]{120}$ **is between 4 and 5.**

Real numbers include both rational and irrational numbers.

Real Numbers

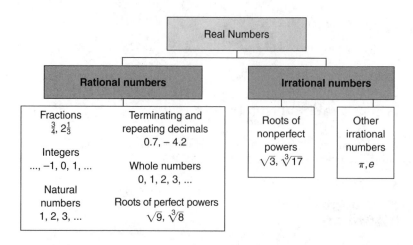

● **RELATED TOPICS:** π, e

How to

Position an irrational number on the number line:

1. Find the perfect powers that the given number is between.
2. Determine which perfect power the given number is nearest.
3. Take the root of each perfect power found in Step 1.
4. Place the root of the given number on the number line between the roots of the perfect powers and nearest the root of the perfect power identified in Step 2.

EXAMPLE   To position $\sqrt{11}$ on the number line:

1, 4, 9, 16   List the perfect squares until you have a number greater than 11.

11 is between 9 and 16.
11 is nearest 9. $11 - 9 = 2$, $16 - 11 = 5$.

$\sqrt{11}$ **is between** $\sqrt{9}$ **and** $\sqrt{16}$ **or 3 and 4. It is closer to 3.**

$\sqrt{9} = 3$, $\sqrt{16} = 4$

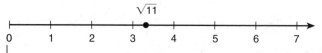

CHAPTER 6

Percents

- ■ **6–1 Percents**
- ■ **6–2 Percentages**

■ 6–1 Percents

The word **percent** means "per hundred" or "for every hundred." Thus, 35 percent means 35 per hundred, or 35 out of every hundred, or $\frac{35}{100}$, and 100 percent means 100 out of 100 parts, or $\frac{100}{100}$, or 1 whole quantity. The symbol % is used to represent "percent."

How to

Change any number to its percent equivalent:

Multiply by 1 in the form of 100%.

EXAMPLE To change the fractions to percent equivalents: $\frac{1}{4}$, $\frac{3}{8}$, $\frac{1}{3}$, $\frac{1}{200}$:

$$\frac{1}{4} \times 100\% = \frac{1}{\underset{1}{4}} \times \frac{\overset{25}{100\%}}{1} = 25\%$$

$$\frac{3}{8} \times 100\% = \frac{3}{\underset{2}{8}} \times \frac{\overset{25}{100\%}}{1} = \frac{75\%}{2} = 37\frac{1}{2}\% \text{ or } 37.5\%$$

$$\frac{1}{3} \times 100\% = \frac{1}{3} \times \frac{100\%}{1} = \frac{100\%}{3} = 33\frac{1}{3}\%$$

$$\frac{1}{200} \times 100\% = \frac{1}{\underset{2}{200}} \times \frac{\overset{1}{100\%}}{1} = \frac{1}{2}\% \text{ or } 0.5\%$$ $\frac{1}{2}\%$ means $\frac{1}{2}$ of every hundredth, or $\frac{1}{2}$ of 1%.

Percents

89

EXAMPLE To change the decimals 0.3 and 0.006 to percent equivalents:

$0.3 \times 100\% = 030.\% = \mathbf{30\%}$ Use the shortcut procedure to
multiply by 100: move the decimal
point two places to the right.

$0.006 \times 100\% = 000.6\% = \mathbf{0.6\%}$ 0.6% means 0.6 of every
hundredth, or 0.6 of 1%.

EXAMPLE To change the whole numbers 1, 3, and 7 to their percent
equivalents:

$1 \times 100\% = \mathbf{100\%}$ 100 out of 100 or all of 1 quantity.

$3 \times 100\% = \mathbf{300\%}$ 300 out of 100 or 3 whole quantities.

More than 100% is more than one whole quantity. 300% is three whole
quantities.

$7 \times 100\% = \mathbf{700\%}$ 7 whole quantities, or 7 times a quantity.

EXAMPLE To change the mixed number $1\frac{1}{4}$ and decimal 5.3 to their percent equivalents:

$$1\frac{1}{4} \times 100\% = \frac{5}{4} \times \frac{\overset{25}{\cancel{100}}\%}{1} = \mathbf{125\%}$$

$$5.3 \times 100\% = 530.\% = \mathbf{530\%}$$

● **RELATED TOPICS:** Multiply whole numbers, fractions,
decimals, and mixed numbers

How to

Change any percent to a numerical equivalent:

Divide by 1 in the form of 100%.

TIP

Division expressed as multiplication

As with fractions, we see that it is convenient to change division to an equivalent multiplication. Is dividing by 100% the same as multiplying by $\dfrac{1}{100\%}$? Yes.

$$A \text{ percent} \div 100\% = A \text{ percent} \div \dfrac{100\%}{1} = A \text{ percent} \times \dfrac{1}{100\%}$$

EXAMPLE   To change the percents 38%, $33\frac{1}{3}\%$, and 5.25% to their fraction and decimal equivalents:

Fraction equivalent

$$38\% = 38\% \div 100\%$$

$$\dfrac{\overset{19}{\cancel{38}}\%}{1} \times \dfrac{1}{\underset{50}{\cancel{100}}\%} = \dfrac{19}{50}$$

$$33\frac{1}{3}\% \div 100\%$$

$$\dfrac{\overset{1}{\cancel{100}}\%}{3} \times \dfrac{1}{\underset{1}{\cancel{100}}\%} = \dfrac{1}{3}$$

Decimal equivalent

$$38\% \div 100\%$$

$$0.38 = 0.38$$

$$33\frac{1}{3}\% \div 100\%$$

$$0.33\frac{1}{3} = 0.33\frac{1}{3}$$

For decimals use the shortcut procedure for dividing by 100%: move the decimal point two places to the left.

First, write the percent in fraction form.

$$5.25\% = 5\dfrac{25}{100}\% = 5\dfrac{1}{4}\%$$

$$5\dfrac{1}{4}\% = 5\dfrac{1}{4}\% \div 100\%$$

$$= \dfrac{21}{4}\% \times \dfrac{1}{100\%}$$

$$= \dfrac{21}{400}$$

$$5.25\% = 5.25\% \div 100\% = 0.0525$$

● **RELATED TOPICS: Dividing fractions and decimals, Reducing Fractions**

Percents

What happens to the % (percent) sign?

From multiplying fractions we can reduce or cancel common factors from a numerator to a denominator. Percent signs and other types of labels can also cancel.

$$\frac{\%}{1} \times \frac{1}{\%} = 1$$

Common Percent, Fraction, and Decimal Equivalents

Percent	Fraction	Decimal	Percent	Fraction	Decimal
10%	$\frac{1}{10}$	0.1	60%	$\frac{3}{5}$	0.6
20%	$\frac{1}{5}$	0.2	$66\frac{2}{3}\%$	$\frac{2}{3}$	$0.66\frac{2}{3}$ or 0.667[a]
25%	$\frac{1}{4}$	0.25	70%	$\frac{7}{10}$	0.7
30%	$\frac{3}{10}$	0.3	75%	$\frac{3}{4}$	0.75
$33\frac{1}{3}\%$	$\frac{1}{3}$	$0.33\frac{1}{3}$ or 0.333[a]	80%	$\frac{4}{5}$	0.8
40%	$\frac{2}{5}$	0.4	90%	$\frac{9}{10}$	0.9
50%	$\frac{1}{2}$	0.5	100%	$\frac{1}{1}$	1.0

[a]These decimals can be expressed with fractions or rounded decimals. The fraction equivalents are exact amounts, and the rounded decimals are approximate amounts.

■ 6–2 Percentages

How to

Find the percent of a number:

1. Change the percent to a numerical equivalent.
2. Multiply the numerical equivalent times the number.

EXAMPLE To find 25% of 124:

$25\% \div 100\% =$ Change the percent to its numerical equivalent.
Write division as equivalent multiplication.

$25\% \times \dfrac{1}{100\%} =$ Perform multiplication and reduce fraction.

$\dfrac{25}{100} = \dfrac{1}{4}$ Multiply the numerical equivalent by the number.

$\dfrac{1}{4} \times \overset{31}{1\!\!\!/\!2\!\!\!/4} = 31$

EXAMPLE To find 12% of $15.24:

$12\% \div 100\% = 0.12$ Change the percent to its numerical equivalent.

$0.12 \times 15.24 =$ Multiply the numerical equivalent times the number.

1.8188 is $1.82 rounded to the nearest cent.

● **RELATED TOPICS: Dividing and multiplying fractions, Rounding decimals**

▲ **APPLICATION: Sales Tax on a Purchase**

Most states, counties, and cities add to the purchase price of many items a certain percent for **sales tax.** Then the **new amount** or **total price** that a purchaser pays is the price of the item plus the sales tax.

How to

Find the sales tax on a purchase:

1. Change the percent (**sales tax rate**) to a numerical equivalent.
2. Multiply the numerical equivalent of the percent times the purchase price.
3. Round to the nearest cent.

EXAMPLE A 5% sales tax is levied on an order of building supplies costing $127.32. To find the amount of sales tax to be paid:

Find 5% of $127.32.

Estimation   Tax: 10% of $127.32 = $12.73 (Move decimal one place to the left.)

5% would be $\frac{1}{2}$ of 10%. $\frac{1}{2}$ of $12.73 is more than $6.

Calculations   5% = 0.05   Change the percent to a numerical
                                          equivalent.

                  0.05 × 127.32 =   Multiply the numerical equivalent times
                                          the purchase price.

                  $6.366 =   Round to the nearest cent.

                  $6.37

L

▲ **APPLICATION: Total Price of a Purchase**

How to

Find the total price of a purchase:

1. Find the sales tax on the purchase.
2. Add the amount of the purchase and the sales tax.

EXAMPLE   To find the total price of an item that costs $46.83 and has a 6% sales tax rate:

    6% = 0.06   Numerical equivalent of 6%.

    0.06 × $46.83 =   Multiply.

    $2.8098 =   Round.

    $2.81   Sales tax.

    $46.83 + $2.81 =   Add purchase price and sales tax.

    $49.64   Total price.

L

▲ **APPLICATION: New Amount or Total Price Found Directly**

How to

Find the new amount (total price) directly:

1. Add 100% and the sales tax rate.
2. Find the numerical equivalent of the new percent.
3. Multiply the numerical equivalent of the new percent times the
   purchase price.
4. Round to the nearest cent.

EXAMPLE To find the total purchase price of an item that costs $32.95 and is subject to a 7% sales tax rate:

100% + 7% =   Add 100% and the sales tax rate.

107%

107% ÷ 100% =   Find the numerical equivalent.

1.07

1.07 × $32.95 =   Multiply the numerical equivalent times the
                     purchase price.

$35.2565   Round to the nearest cent.

$35.26   Total price.

▲ APPLICATION: Commission on Sales

Sales people are sometimes paid based on the amount of sales made. This method of payment is called selling **on commission.**

How to

Find the commission on sales:

1. Change the **rate of commission** (percent) to a numerical equivalent.
2. Multiply the numerical equivalent times the total sales.
3. Round to the nearest cent.

EXAMPLE A salesperson receives a 6% commission on all sales. If this salesperson sells $15,575 in merchandise during a given pay period, the commission is:

Estimation   10% of $15,575 = $1,557.50   Move decimal one place to
                                              the left.

$$\frac{1}{2} \text{ of } 10\% = 5\%$$

$$5\% \text{ of } \$15,575 = \frac{1}{2} \text{ of } 10\% \text{ of } \$15,575 \approx \$750$$

                                  ≈ is read "is approximately."

6% commission > $750   Since 6% > 5%.

Calculations   6% = 0.06   Numerical equivalent of
                                    commission rate.

$$0.06 \times 15{,}575 =$$ Multiply the numerical
equivalent times the total sales.

$934.50   Commission.

Estimating percents

Estimate percents by comparing them to a percent that you can calculate mentally.

To find

1% of a number	*Multiply by 0.01*	Move decimal
2% of a number	2 × 1% of a number	two places to
3% of a number	3 × 1% of a number	left.
10% of a number	*Multiply by 0.1*	Move decimal
5% of a number	1/2 of 10% of a number	one place to
20% of a number	2 × 10% of a number	left.
30% of a number	3 × 10% of a number	
50% of a number	*1/2 of a number*	Divide by 2.
25% of a number	*1/4 of a number*	Divide by 4.
75% of a number	3 × 25% of a number	
33 1/3% of a number	*1/3 of a number*	Divide by 3.
66 2/3% of a number	2 × 33 1/3% of a number	

EXAMPLE To find 10% of $51.00:

10% of $51.00 = **$5.10.**   Mentally, move the decimal point in the
number one place to the left.

▲ **APPLICATION: Estimate a 15% Tip**

How to

Estimate a 15% tip:

1. Find 10% of the total bill.
2. Find 5% of the total bill by taking $\frac{1}{2}$ of the 10% amount.
3. Add the amounts for 10% and 5% of the bill.

EXAMPLE   To estimate a 15% tip on a restaurant bill of $47.50:

10% of 47.50 = $4.75   Find 10% of the total bill.

5% of 47.50 = $\frac{1}{2}$ of 4.75 = $2.38   Find 5% of the total bill.

15% = 10% + 5%

$4.75 + $2.38 =   Add the amounts for 10% and 5%.

$7.13   15% tip.

└─

▲ **APPLICATION: Estimate a 20% Tip**

How to

Estimate a 20% tip:

1. Find 10% of the total bill.
2. Double the 10% amount.

EXAMPLE   A taxi fare is $23.75. To find the amount of a 20% tip:

10% of $23.75 = $2.375   Find 10% of the fare.

2 × $2.375 = $4.75   Double the 10% amount.

20% tip is **$4.75**

└─

▲ **APPLICATION: Discount**

Retail stores frequently reduce the price of merchandise from the **original price**. The amount by which an item is reduced is called a **discount** or **mark-down**. The new price is called the **reduced price** or **sale price**.

How to

Find the discount of an item:

1. Change the percent of discount to a numerical equivalent.
2. Multiply the numerical equivalent times the original price.
3. Round to the nearest cent.

Percents

EXAMPLE To find a 30% discount on a book that costs $49:

30% = 0.3   Change the percent to a numerical equivalent.

0.3 × 49 =   Multiply the numerical equivalent times the original price.

$14.70   Discount amount.

└─

▲ **APPLICATION: Sale or Reduced Price**

How to

Find the sale or reduced price of an item:

1. Find the discount amount for the item.
2. Subtract the discount amount from the original price.

EXAMPLE To find the reduced price of an item that costs $49 and is reduced 30%:

30% = 0.3

0.3 × 49 = $14.70   Find the discount amount.

$49.00 − $14.70 =   Subtract the discount amount from the original price.

$34.30   Sale or reduced price.

└─

▲ **APPLICATION: Sale Price or Reduced Price Directly**

How to

Find the reduced price directly:

1. Subtract the discount rate from 100%.
2. Find the numerical equivalent of the new percent.
3. Multiply the numerical equivalent of the new percent times the original price.
4. Round to the nearest cent.

EXAMPLE   To find the reduced price directly of an item that costs $49 and is reduced 30%:

100% − 30% = 70%   Subtract the discount rate from 100%.

70% = 0.7   Find the numerical equivalent of the new percent.

0.7 × $49.00 =   Multiply the numerical equivalent times the original price.

$34.30   Reduced or sale price.

It is helpful in developing our number sense with percents to think of percents in pairs. 100% is one whole quantity. A percent that is less than 100% represents part of a quantity. For example, 30% represents part of one quantity. Then, 70% is the rest of the quantity if 30% is removed. This percent, 70%, is the complement of 30%. The **complement of a percent** is the difference between 100% and a given percent.

How to

Find the complement of a percent:

Subtract the percent from 100%.

EXAMPLE   To find the complements of (a) 25%, (b) 80%, and (c) 36%:

(a)   100% − 25% = **75%**   Subtract from 100%.

(b)   100% − 80% = **20%**   Subtract from 100%.

(c)   100% − 36% = **64%**   Subtract from 100%.

▲ **APPLICATION: Estimate the Amount You Pay for a Sale Item**

How to

Estimate the amount you pay for a sale item:

1. Round the original price and the percent to numbers you can work with mentally.
2. Find the complement of the rounded percent.

3. Relate the complement to 10% by dividing it by 10%.
4. Find 10% of the rounded original price.
5. Multiply the results from steps 3 and 4.

EXAMPLE To estimate the amount you pay on a $49.99 item advertised at 70% off:

$49.99 rounds to $50
100% − 70% = 30%   Complement of 70% (percent you pay).

30% ÷ 10% = 3   Relate to 10%.

10% of $50 = $5   Move decimal one place to the left.

3 × $5 = **$15**   Three times 10% of $50.

EXAMPLE To estimate the amount you pay on a $28 item advertised at 18% off.

18% rounds to 20%.
$28 rounds to $30.
100% − 20% = 80%   Complement of 20% (percent you pay).

80% ÷ 10% = 8   Relate to 10%.

10% of $30 = $3   Move decimal one place to the left.

8 × $3 = **$24**   Eight times 10% of $30.

Equations and Formulas

CHAPTER 7

Linear Equations

- **7–1 Variable Notation**
- **7–2 Solving Linear Equations in One Variable**
- **7–3 Solving Linear Equations with Fractions and Decimals**
- **7–4 Equations Containing One Absolute-Value Term**
- **7–5 Formula Evaluation and Rearrangement**

■ 7–1 Variable Notation

Symbols allow mathematical expressions to be written in a concise form that overcomes many language and cultural barriers. For symbols to be effective, there must be a widespread understanding of the numbers and concepts represented by the symbols. A letter that represents an unknown or missing number is called a **variable** or **unknown**. A number is called a **constant**. Numbers or variables that are multiplied are called **factors.**

Multiplication notation conventions

When a term is the product of letters alone (such as *ab* or *xyz*) or the product of a number and one or more letters (such as 3*x* or 2*ab*), parentheses or other symbols of multiplication are usually omitted. Thus, *ab* means *a* times *b*, 3*x* means 3 times *x*, and so on.

Algebraic expressions that are single quantities or quantities that are added or subtracted are called **terms.** A plus or minus sign that is not within a grouping separates terms.

> ### How to
>
> **Identify terms:**
>
> 1. Separate an expression into terms.
> 2. The sign of each term is the sign that precedes the term.
>
> $$3a + b \qquad\qquad 3(a + b) \qquad\qquad \frac{3}{a + b}$$
>
> two terms   one term   one term

EXAMPLE To identify the terms in each expression, draw a box around each term:

(a) $3x + 5$ (b) $2ab + 4a - b + 2(a + b)$

(c) $\dfrac{2a + 1}{3}$ (d) $\dfrac{2a}{3} + 1$

(a) $\boxed{3x} + \boxed{5}$

Terms are separated by " + " and " − " signs that are not within a grouping. The expression has two terms.

(b) $\boxed{2ab} + \boxed{4a} - \boxed{b} + \boxed{2(a + b)}$

The plus sign within the grouping does not separate terms.

(c) $\dfrac{\boxed{2a + 1}}{3}$

The fraction bar is a grouping symbol. The numerator is a grouping, $2a + 1$. The expression has one term.

(d) $\boxed{\dfrac{2a}{3}} + \boxed{1}$

The expression has two terms.

A term that contains only numbers is a **number term** or **constant.** A term that contains only one letter, several letters used as factors, or a combination of letters and numbers used as factors is a **letter term** or **variable term.**

The numerical factor of a term is the **numerical coefficient.** Unless otherwise specified, we will use **coefficient** to mean the *numerical coefficient.* The numerical factor should be written *in front* of the variable.

How to

Identify the numerical coefficient of a term:

1. Since a term contains only factors, the coefficient is the number factor or the product of all number factors.
2. If the term is a fraction (an indicated division),
   a. Write the number in the denominator (indicated division) as an equivalent multiplication.
   b. Write the product of all numerical factors.
   c. The numerical coefficient is the product from Step 2b.

EXAMPLE   To identify the numerical coefficient in each term:

(a) $2x$ (b) $-3ab$ (c) $4(x + 3)$ (d) $-\dfrac{n}{3}$ (e) $\dfrac{2b}{5}$

(a)   The coefficient of $2x$ is **2**.

(b)   The coefficient of $-3ab$ is **−3**.

(c)   The coefficient of $4(x + 3)$ is **4**.

(d) $-\dfrac{n}{3}$ is the same as $-\dfrac{1}{3}n$, so the coefficient is $-\dfrac{1}{3}$.

(e) $\dfrac{2b}{5} = \dfrac{1}{5}(2b) = \dfrac{1}{5}(2)b = \dfrac{2}{5}b$. The coefficient is $\dfrac{2}{5}$.

TIP

Coefficient of 1 or −1

When a letter term has no written numerical coefficient, the coefficient is understood to be 1, so $x = 1x$. Similarly, the numerical coefficient of $-x$ is −1, so $-x = -1x$.

An **equation** is a symbolic statement that two expressions are equal. An equation may be true or false. To **solve** an equation is to find values of the missing numbers that make the equation true. Letters such as a, x, or z are used to represent the missing number or numbers in an equation.

The **root** or **solution** of an equation is the value of the variable that makes the equation true.

Three properties of equalities are used in a variety of circumstances in mathematics.

Reflexive property of equality:

A number or expression equals itself: $a = a$

Symmetric property of equality:

If the sides of an equation are interchanged, equality is maintained. If $a = b$, then $b = a$.

Transitive property of equality:

If one expression equals a second expression and the second expression equals a third expression, the first expression equals the third expression. If $a = b$ and $b = c$, then $a = c$.

Symbolic representations of written statements or real-life situations allow us to more systematically determine the value of missing amounts.

How to

Translate statements into equations:

1. Assign a letter to represent the missing number.
2. Identify key words or phrases that imply or suggest specific operations.
3. Translate words into symbols.

EXAMPLE   To translate the statements into symbols:

(a)   The sum of 12, 23, and a third number is 52.
The third number is missing.
Let x represent the third number.

$12 + 23 + x = 52$ Translate the entire statement. *Sum* indicates addition. *Is* translates to *equals.*

(b)   When a number is subtracted from 45, the result is 17.
The number being subtracted (the subtrahend or second number) is missing.
Let x represent the missing number.

$45 - x = 17$ Translate the entire statement.

Linear
Equations

Addition:	the sum of, plus, increased by, more than, added to, exceeds, longer, total, heavier, older, wider, taller, gain, greater than, more, expands
Subtraction:	less than, decreased by, subtracted from, the difference between, diminished by, take away, reduced by, less, minus, shrinks, younger, lower, shorter, narrower, slower, loss
Multiplication:	times, multiply, of, the product of, multiplied by
Division:	divide, divided by, divided into, how big is each part, how many parts can be made from

EXAMPLE   To translate the statements into symbols:
How many shelves that are each 3 feet long can be made from a board that is 12 feet long?

> The missing number is the number of shelves that can be made from one 12-ft board.
> Let x represent the number of shelves. The number of shelves, x times the length of each shelf, 3, equals 12.
> How many shelves that are 3 feet long can be made from a board that is 12 feet long? $3x = 12$ or $\frac{12}{3} = x$.

Like terms are number terms or letter terms that have exactly the same letters.

How to

Simplify variable expressions:

1. Change subtraction to addition if appropriate.
2. Add the numbers using the appropriate rule for adding signed numbers. The sum is a signed number.
3. Add the numerical coefficients of the like variables using the appropriate rule for adding signed numbers. The sum has the same letter factor or factors as the like terms being added.

EXAMPLE   To simplify the expressions, combine like terms:

(a) $5a + 2a - a$ (b) $3x + 5y + 8 - 2x + y - 12$

(a) $5a + 2a - a = 6a$ Add coefficients of a:
                                  $5 + 2 - 1 = 6$.

(b) $3x + 5y + 8 - 2x + y - 12$ Add like terms.

 $= x + 6y - 4$ $3x - 2x = x$
                         $5y + y = 6y$
                         $8 - 12 = -4$

■ 7–2 Solving Linear Equations in One Variable

A **basic equation** is an equation that consists of only one variable and its coefficient on one side of the equal sign and only a constant (number) term on
the other side.

An equation is **solved** when the letter is alone with a coefficient of $+1$
on one side of the equation and a number is on the other side. The number
on the side opposite the letter is called the **root** or **solution** of the equation.
The root or solution is the number that makes a true statement when put in
place of the variable in the original equation.

Both sides of an equation may be multiplied by the same *nonzero* quantity without changing the equality of the two sides. Symbolically, if $a = b$
and $c \neq 0$, then $ac = bc$. This concept is called the **multiplication property
of equality** or the **multiplication axiom**.

How to

Solve a basic equation:

1. Divide both sides of the equation by the numerical coefficient of
   the letter term; or
2. Multiply both sides of the equation by the reciprocal of the coefficient of the letter term.

EXAMPLE   To solve the equation $4x = 48$:

$\dfrac{4x}{4} = \dfrac{48}{4}$ Divide both sides of the equation by the coefficient of $4x$ or 4.

$x = 12$ Solution or root.

Linear
Equations

EXAMPLE To solve the equation $\frac{1}{4}n = 7$:

Using multiplication: $\frac{1}{4}n = 7$

$$\left(\frac{4}{1}\right)\frac{1}{4}n = 7\left(\frac{4}{1}\right)$$ Multiply by the reciprocal of the coefficient of x.

$$n = 28$$ The reciprocal of $\frac{1}{4}$ is $\frac{4}{1}$.

Using division: $\frac{1}{4}n = 7$

$$\frac{\frac{1}{4}n}{\frac{1}{4}} = \frac{7}{\frac{1}{4}}$$ Divide by the coefficient of x. $7 \div \frac{1}{4} = \frac{7}{1} \times \frac{4}{1} = 28$

$$n = 28$$

TIP

Dividing by the coefficient versus multiplying by the reciprocal of the coefficient

When the equation contains only *whole numbers* or *decimals*, it is generally more convenient to divide by the coefficient of the variable or letter term than to multiply by the reciprocal of the coefficient of that variable. When the equation contains a *fraction*, it is often preferable to multiply by the reciprocal of the coefficient of the variable.

● **RELATED TOPICS: Dividing fractions, Multiplying fractions, Reciprocal**

EXAMPLE To solve the equation $-n = 25$:
The coefficient is -1, so the equation is not solved. The coefficient must be $+1$ for the equation to be solved.

$$\frac{-1n}{-1} = \frac{25}{-1}$$ Divide by the coefficient of the letter term.

$$n = -25$$

On which side should the letter be?

Multiplication and division are effective regardless of which side of the equation contains the letter term. Once the equation is solved, it is preferred that the letter be on the left ($x = 90$ instead of $90 = x$). Either way is correct.

How to

Check or verify the solution or root of an equation:

1. Substitute the solution in place of the letter each time it appears in the original equation.
2. Perform all indicated operations on each side of the equation using the order of operations.
3. If the solution is correct, the value of the left side of the equation should equal the value of the right side.

EXAMPLE   In a previous example the solution for the equation $4x = 48$ was found to be 12. Check or verify this root.

$4x = 48$ Substitute 12 for x.

$4(12) = 48$ Perform the multiplication.

$48 = 48$ The root is verified if the final equality is true.

How to

Solve equations with like terms on the same side of the equation:

1. Combine (add or subtract) like terms that are on the same side of the equal sign.
2. Multiply both sides of the equation by the reciprocal or divide both sides of the equation by the coefficient of the variable.

Linear
Equations

EXAMPLE   To solve $2x + 3x = 13 + 2$ for x:

$$2x + 3x = 13 + 2$$

Combine like terms on each side of the equal sign.

$$5x = 15$$

Divide both sides of the equation by 5, the coefficient of x.

$$\frac{5x}{5} = \frac{15}{5}$$

$$x = 3$$

The same quantity can be added to both sides of an equation without changing the equality of the two sides. Symbolically, if $a = b$, then $a + c = b + c$. This concept is called the **addition property of equality** or the **addition axiom**.

How to

Solve equations with like terms on opposite sides of the equation:

1. Identify the like terms and determine which term or terms should be moved to create a basic equation.
2. Add to both sides of the equation the opposite of the term that is to be moved.
3. Combine like terms on each side of the equation.
4. Solve the basic equation.

EXAMPLE   To solve $9 - 4x = 8x$:

$$9 - 4x = 8x$$

Add $4x$, the opposite of $-4x$, to both sides. (Addition property)

$$9 \; -4x + 4x = 8x + 4x$$

Combine like terms on both sides. $-4x + 4x = 0$, $8x + 4x = 12x$

$$9 + 0 = 12x$$

Combine like terms on left. $9 + 0 = 9$

$$9 = 12x$$

$$\frac{9}{12} = \frac{12x}{12}$$

Divide both sides by the coefficient of x to solve the basic equation. (Multiplication property)

$$\frac{9}{12} = x$$

Reduce fraction to lowest terms.

$$\frac{3}{4} = x \quad \text{or} \quad x = \frac{3}{4}$$

TIP

Sorting or transposing terms

Transposing is a shortcut for moving terms using the addition property or axiom. Add the opposite of a term to both sides of an equation. That is, *omit the term from one side of the equation and write its opposite on the other side.* This process is also described as *sorting* the terms so that like terms are on the same side of the equation. We are *mentally* adding the opposite term to both sides.

Distributive Property. A multiplication of a factor times a sum or difference of two terms can be changed to addition of terms by multiplying the first factor by each term of the second factor: $a(b + c) = ab + ac$.

● **RELATED TOPICS: Signed numbers, Whole numbers**

How to

Solve equations that contain parentheses:

1. Apply the distributive property to *remove parentheses.*
2. *Combine like terms* on each side of the equation.
3. *Sort terms* to collect the variable or letter terms on one side and constant or number terms on the other.
4. *Combine like terms* on each side of the equation.
5. *Solve the basic equation* by multiplying by the reciprocal of the coefficient of the variable term or dividing by the coefficient of the variable term.

EXAMPLE   To solve and check the equation $28 = 7x - 3(x - 4)$ for x:

$28 = 7x \;\; - 3(x - 4)$ Each term in the parentheses is multiplied by -3 using the distributive property.

$28 = 7x - 3x + 12$ Combine like terms on the right.

$28 = 4x + \;\; 12$ Sort terms.

Linear Equations

$$28 - 12 = 4x$$ Combine like terms on the left.

$$16 = 4x$$ Divide to solve the basic equation.

$$\frac{16}{4} = \frac{4x}{4}$$

$$4 = x$$

As equations get more involved, the importance of checking becomes more apparent. To check the root 4, we substitute 4 for x in the equation.

$$28 = 7x - 3(x - 4)$$

$$28 = 7(4) - 3(4 - 4)$$ Substitute 4 for x. Then simplify the grouping $4 - 4 = 0$.

$$28 = 7(4) - 3(0)$$ Multiply.

$$28 = 28 - 0$$ Subtract.

$$28 = 28$$

■ 7–3 Solving Linear Equations with Fractions and Decimals

How to

Solve an equation containing fractions:

1. Multiply each term of the *entire* equation by the least common multiple (LCM) of the denominators of the equation.
2. Apply the distributive property to remove parentheses.
3. Combine like terms on each side of the equation.
4. Sort terms to collect the variable or letter terms on one side and number terms on the other.
5. Combine like terms on each side of the equation.
6. Solve the basic equation by multiplying by the reciprocal of the co-efficient of the variable term or dividing by the coefficient of the variable term.

EXAMPLE   To solve $-\dfrac{1}{4}x = 9 - \dfrac{2}{3}x$ by clearing all fractions first:

$$-\frac{1}{4}x = 9 - \frac{2}{3}x$$ Multiply each term by the LCM of the denominators. Reduce. LCM = 12 or 4(3).

$$(\;4\;)(\;3\;)\left(-\frac{1}{4}x\right) = (\;4\;)(\;3\;)(9) - (\;4\;)(\;3\;)\left(\frac{2}{3}x\right)$$

$$(\overset{1}{\cancel{4}})(3)\left(-\frac{1}{4}x\right) = (4)(3)(9) - (4)(\overset{1}{\cancel{3}})\left(\frac{2}{3}x\right)$$ Multiply the remaining factors.

$$-3x = 108 - 8x$$ This equation contains no fractions. Sort terms.

$$-3x + 8x = 108$$ Combine like terms.

$$5x = 108$$ Divide by the coefficient of x.

$$\frac{5x}{5} = \frac{108}{5}$$

$$x = \frac{108}{5}$$

▲ APPLICATION: Amount of Work and Rate of Work

A **rate measure** involves a unit of time. If car A travels 50 miles in 1 hour, then car A's **rate of work** (travel) is 50 miles per 1 hour, or $\frac{50 \text{ mi}}{1 \text{ hr}}$ expressed as a fraction. If car A travels for 3 hours, then the **amount of work** is $\frac{50 \text{ mi}}{\text{hr}} \times 3 \text{ hr} = 150$ miles.

● RELATED TOPIC: U.S. customary rate measure

How to

Find the amount of work produced by one individual or machine:

1. Identify the rate of work and the time worked.
2. Use the formula for amount of work.

Formula for amount of work:

Amount of work completed = Rate of work × Time worked

Linear Equations

EXAMPLE A carpenter can install 1 door in 3 hours. To find the number of doors the carpenter can install in 30 hours:

Known facts   Rate of work = 1 door per 3 hours, $\frac{1}{3}$ door per hour, or
$\frac{1 \text{ door}}{3 \text{ hr}}$

Time worked = 30 hours.

Unknown facts W = Amount of work or number of doors installed

Relationships   Amount of work = Rate of work × Time worked

Estimation   Several doors can be installed in 30 hr., but less than 30 since 1 door would have to be installed every hour to install 30 doors.

Calculations $W = \dfrac{1 \text{ door}}{\overset{}{3} \text{ hr}} \times \overset{10}{30} \text{ hr}$ Reduce and multiply.

$W = 10$ doors

Interpretation   **Thus, 10 doors can be installed in 30 hours.**

⌐

● **RELATED TOPIC: Direct measurement**

▲ **APPLICATION: Work Accomplished
by Two or More Together**

If two workers or machines do a job together, we can find the amount of work done by each worker or machine. Combined, the amounts equal 1 total job.

How to

*Find the amount of work each individual or machine
produces when working together:*

1. Identify the rate of work for each individual or machine. If unknown, assign a letter to represent the unknown.
2. Identify the time worked for each individual or machine. If unknown, assign a letter to represent the unknown.
   Note: Only one letter should be used and other unknowns should be written in relationship to the one letter.
3. Use the formula for completing one job.

*Formula for completing one job when A and B are
working together:*

$$\left(\begin{array}{c} \text{A's} \\ \text{amount of} \\ \text{work} \end{array} \right) + \left(\begin{array}{c} \text{B's} \\ \text{amount of} \\ \text{work} \end{array} \right) = 1 \text{ completed job}$$

or

$$\begin{pmatrix} \text{A's} \\ \text{rate of} \quad \times \quad \text{time} \\ \text{work} \qquad \text{worked} \end{pmatrix} + \begin{pmatrix} \text{B's} \\ \text{rate of} \quad \times \quad \text{time} \\ \text{work} \qquad \text{worked} \end{pmatrix} = 1 \text{ completed job}$$

EXAMPLE   Pipe 1 fills a tank in 6 min and pipe 2 fills the same tank in 8 min. To find how long it takes for both pipes together to fill the tank:

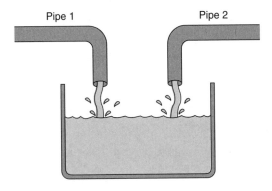

Pipe 1   Pipe 2

Known facts   Pipe 1 fills the tank at a rate of 1 tank per 6 min, $\frac{1}{6}$ tank per minute, or $\frac{1 \text{ tank}}{6 \text{ min}}$.

Pipe 2 fills the tank at a rate of 1 tank per 8 min, $\frac{1}{8}$ tank per minute, or $\frac{1 \text{ tank}}{8 \text{ min}}$.

Unknown facts T = time (in minutes) for both pipes together to fill the tank.

Relationships   Amount of work of pipe 1 = $\frac{1 \text{ tank}}{6 \text{ min}}$ (T). Amount of work of pipe 2 = $\frac{1 \text{ tank}}{8 \text{ min}}$ (T).

Amount of work together = Pipe 1's work + Pipe 2's work.

Estimation   Both pipes together should fill the tank in less time than the pipe with the slower rate, or in less than 6 min.

Calculations $$\frac{1 \text{ tank}}{6 \text{ min}} (T \text{ min}) + \frac{1 \text{ tank}}{8 \text{ min}}(T \text{ min}) = 1 \text{ tank}$$

The LCM is 24 .

$$(24) \left(\frac{1}{6} T \right) + (24) \left(\frac{1}{8} T \right) = (24)(1)$$

Clear the fractions by multiplying by the LCM.

$$(\overset{4}{24}) \left(\frac{1}{6} T \right) + (\overset{3}{24}) \left(\frac{1}{8} T \right) = (24)(1)$$

Reduce and multiply.

$$4T + 3T = 24$$

Combine like terms.

$$7T = 24$$

$$\frac{7T}{7} = \frac{24}{7}$$

Divide by the coefficient of T.

$$T = \frac{24}{7}\left(\text{or } 3\frac{3}{7}\right)\text{min}$$

Interpretation   **Both pipes together fill the tank in $3\frac{3}{7}$ min.**

How to

Solve a decimal equation by clearing decimals:

1. Multiply each term of the *entire* equation by the least common multiple (LCM) of the fractional amounts following the decimal points.
2. Follow the same steps used in solving a linear equation.

TIP

LCM for decimals

Digits to the right of the decimal point represent fractions whose denominators are determined by the place value. To find the LCM for all the decimal numbers in an equation, *look* at the decimal with the most digits after the decimal point. Use its denominator to clear the decimals.

This procedure allows you to avoid dividing by a decimal, which can be a common source of error when making calculations by hand.

EXAMPLE   To solve $0.38 + 1.1y = 0.6$ by first clearing the equation of decimals:

$$0.38 + 1.1y = 0.6$$ The LCM is 100.

$$100\,(0.38) + 100\,(1.1y) = 100\,(0.6)$$ Multiply by 100.

$$38 + 110y = 60$$ Sort terms.

$$110y = 60 - 38 \qquad \text{Combine like terms.}$$

$$\frac{110y}{110} = \frac{22}{110} \qquad \text{Divide by the coefficient of } y.$$

$$y = 0.2$$

$\llcorner$

▲ APPLICATION: Percentage Formulas

The **percentage formula,** Percentage = Rate × Base, can be written as $P = R \times B$. The letters or words represent numbers. When the numbers are put in place of the letters, the formula guides you through the calculations.

In the formula $P = R \times B$, the **base** (B) represents the original number or entire quantity. The **percentage** (P) represents a **portion** of the base. The **rate** (R) is a percent that tells us how the base and percentage are related. In the statement "50 is 20% of 250," 250 is the base (the entire quantity), 50 is the percentage (part), and 20% is the rate (percent).

The three percentage formulas are

Percentage = Rate × Base	$P = R \times B$	for finding the percentage
Base = $\dfrac{\text{Percentage}}{\text{Rate}}$	$B = \dfrac{P}{R}$	for finding the base
Rate = $\dfrac{\text{Percentage}}{\text{Base}}$	$R = \dfrac{P}{B}$	for finding the rate

Circles can help us visualize these formulas. The shaded part of the circle represents the missing amount. The unshaded parts represent the known amounts. If the unshaded parts are *side by side, multiply* their corresponding numbers to find the missing number.

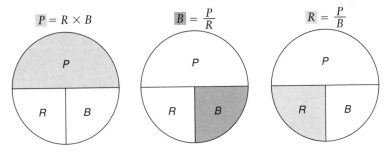

If the unshaded parts are *one on top of the other, divide* the corresponding numbers to find the missing number.

Identify the rate, base, and percentage

The following descriptions may help you recognize the rate, base, or percentage more quickly:

> *Rate* is usually written as a percent, but it may be a decimal or fraction.
> *Base* is the total amount, original amount, entire amount, and so on;
> it is the amount that the *percentage* is a portion of. In a sentence the
> base is often closely associated with the preposition *of*.
> *Percentage* is called part, partial amount, portion, amount of increase or decrease, amount of change, and so on; it is a portion of
> the *base*. In a sentence the percentage is often closely associated with
> a form of the verb *is*.

A common memory jogger for a percent is Percent or Rate $= \dfrac{is}{of}$.

How to

Use the percentage formula to solve percentage problems:

1. Identify and classify the two known values and the one missing value.
2. Choose the appropriate percentage formula for finding the missing value.
3. Substitute the known values into the formula. For the rate use the decimal or fractional equivalent of the percent.
4. Perform the calculation indicated by the formula.
5. Interpret the result. Convert decimal or fractional equivalents of the rate to a percent.

EXAMPLE   To solve the problems (a) 20% of 400 is what number? (b) 20% of what number is 80? (c) 80 is what percent of 400?

(a)   20% = Rate   Identify known values and missing value.

400 = Base

Percentage is missing   Choose the appropriate formula.

$P = R \times B$ Substitute values.

$P = 0.2 \times 400$ Perform calculation.

$P = 80$ Interpret result.

20% of 400 is 80.

(b)   20% = Rate   Identify known values and missing value.

80 = Percentage

Base is missing   Choose the appropriate formula.

$B = \dfrac{P}{R}$ Substitute values.

$B = \dfrac{80}{0.2}$ Perform calculation.

$B = 400$ Interpret result.

20% of 400 is 80.

(c)   80 = Percentage   Identify know values and missing value.

400 = Base

Rate is missing   Choose the appropriate formula.

$R = \dfrac{P}{B}$ Substitute values.

$R = \dfrac{80}{400}$ Perform calculation.

$R = 0.2$ Interpret result. $0.2 = 20\%$

80 is 20% of 400.

▲ APPLICATION: Qualifying for a Home Mortgage

A mortgage company requires that the monthly cost of a new home be no more than 28% of the buyer's gross monthly income. To find the monthly income a prospective buyer must have or exceed to qualify for a home that has a monthly cost (including taxes and insurance) of $1,200:

28% = Rate   Identify the known values and the missing value.

$1,200 = Percentage

Base is missing

$B = \dfrac{P}{R}$

Choose the appropriate formula.

Substitute values.

$B = \dfrac{1,200}{0.28}$

Perform calculation.

$B = 4285.71$

Interpret result.

Prospective buyer must have a monthly income of at least $4286, rounded to the nearest dollar.

▲ APPLICATION: Simple Interest

Whether for a savings account or a loan, the terms **interest, principal, rate,** and **time** are used similarly. Interest is the amount of money earned or paid. Principal is the amount of money invested or borrowed. Rate is the percent of the principal that is earned or paid. The rate is based on some period of time, often 1 year. The time is the length of time involved for the investment or loan.

How to

Find the simple interest for an investment or a loan:

1. Change the rate to a decimal or a fraction.
2. Substitute the known values in the interest formula.
3. Solve the formula (equation) for the missing value.

Formula for simple interest:

$I = PRT$

where I = **interest**, P = **principal**, R = **rate** or percent in decimal form, and T = **time**.

EXAMPLE   A $1,000 investment is made for $2\frac{1}{2}$ years at 8.25% per year. To find the amount of interest:

Rate = 8.25% = 0.0825

Change the interest rate to a decimal equivalent.

$I = PRT$

Substitute the known values in the formula.

principal   rate   time

$I = \$1,000 \times 0.0825 \times 2.5$ years   Time $= 2\frac{1}{2}$ years $= 2.5$ years.

$I = \$206.25$

The interest for $2\frac{1}{2}$ years is $\$206.25$.

▲ APPLICATION: 90 Days Same as Cash! Deals

Just before Christmas, Kwanza, and Hanukkah, stores advertise "90 days—same as cash!" deals on home appliances, furniture, electronics, and so on, with a minimum purchase of $500. The customer must be approved for credit and must sign a contract before purchase agreeing to some specified terms: If the customer pays in full within 90 days, he or she pays only the sales price plus tax. But if payment is not made *in full* within 90 days, 26.8% annual simple interest is added to the sales price plus tax for the first 90 days, plus 2% simple interest per month on the unpaid balance, with minimum payments of $50 per month. A full month's interest is charged even if a portion of a month is used. Finance companies, with their established methods of credit approval, collection, repossession, and litigation, usually handle these contracts for the stores. Research shows that only about 30% of customers pay off these bills within 90 days, and most customers become continuous or repeat customers of the finance company after the initial loan is paid off.

EXAMPLE   On December 15, a family purchases a washer/dryer set under these terms for $699 plus 8.25% sales tax for a total of $756.67. They plan to pay it off within 90 days with the anticipated $1,000 IRS refund. The refund doesn't arrive until April 1. How much does it cost her to pay off this loan 15 days late?

$\$756.67(0.268)\dfrac{90}{365} = \$50.00$ Interest (rounded) for 90 days at 26.8% interest. $I = PRT$.

$\$756.67 + \$50.00 = \$806.67$ New unpaid balance.

$\$806.67(0.02)(1) = \$16.13$ Interest (rounded) on unpaid balance for 1 month.

$\$806.67 + \$16.13 = \$822.80$ Total amount paid.

$\$822.80 - \$756.67 = \mathbf{\$66.13}$ Amount of interest paid.

A **ratio** is a fraction.

A **proportion** is a statement of two equal ratios (or **fractions**). $\dfrac{a}{b} = \dfrac{c}{d}$

(b and $d \neq 0$). A **cross product** of a proportion is the product of the denominator of one fraction and the numerator of the other fraction.

Property of proportions:

The cross products of a proportion are equal.

If $\dfrac{a}{b} = \dfrac{c}{d}$,   then $ad = bc$, $b, d \neq 0$

The property of proportions is stated as: The product of the **extremes** (end factors a and d) equals the product of the **means** (middle factors b and c).

   The **extremes** of a proportion are the numerator of the first fraction and the denominator of the second fraction. The **means** of a proportion are the denominator of the first fraction and the numerator of the second fraction. A proportion can also be written as $a : b = c : d$.

How to

Verify that two ratios form a proportion:

$$\text{Is } \dfrac{4}{6} = \dfrac{12}{18} \text{ a proportion?}$$

1. Find the two cross products. $4 \times 18 = 72$ $6 \times 12 = 72$
2. Compare the two cross products.   Cross products are equal.
3. If the cross products are equal,
   the two ratios form a proportion.   The ratios are proportional.

EXAMPLE   To select two ratios that form a proportion from the set of ratios:

$$\dfrac{1}{2}, \dfrac{3}{8}, \dfrac{6}{12}, \dfrac{5}{15}$$

$\dfrac{1}{2}, \dfrac{3}{8}$ $1 \times 8 = 8, \ 2 \times 3 = 6$ Find the two cross products of a pair of fractions.

$8 \neq 6$ Compare the cross products. Not a proportion.

$\dfrac{1}{2}, \dfrac{6}{12}$ $1 \times 12 = 12, \ 2 \times 6 = 12$ Find the cross products of a different pair of fractions.

$12 = 12$ The two cross products are
                                             equal.

$\dfrac{1}{2} = \dfrac{6}{12}$ The two ratios form a
                                             proportion.

└─

How to

Solve a proportion for any missing element:

1. Cross multiply.
2. Solve the resulting equation.

EXAMPLE   To solve a proportion for x:

$\dfrac{2}{3} = \dfrac{6}{x}$ Cross multiply.

$2x = 3(6)$

$\dfrac{2x}{2} = \dfrac{18}{2}$ Divide by the coefficient of x.

$x = 9$

└─

EXAMPLE   To solve the proportion for x:

$\dfrac{x - 2}{x + 8} = \dfrac{3}{5}$ Cross multiply.

$5(x - 2) = 3(x + 8)$ Distribute.

$5x - 10 = 3x + 24$ Sort terms.

$5x - 3x = 24 + 10$ Combine like terms.

$2x = 34$ Solve the basic equation.

$\dfrac{2x}{2} = \dfrac{34}{2}$

$x = 17$

└─

Linear Equations

▲ APPLICATION: Direct Proportions or Direct Variation

The proportion format is frequently used to solve a wide variety of application problems. A **direct proportion** is one in which the quantities being compared are directly related, so that as one quantity increases (or decreases), the other quantity also increases (or decreases). This relationship is also called a **direct variation.**

How to

Set up and solve a direct proportion:

1. Establish two pairs of related data.
2. Write one pair of data in the numerators of the two ratios.
3. Write the other pair of data in the denominators of the two ratios.
4. Form a proportion using the two ratios.
5. Solve the proportion.

EXAMPLE   To find (a) the cost of 10 apples if 4 apples cost $1 and (b) to determine how many apples can be purchased for $10.

(a)   Pair 1:   4 apples cost $1.

Pair 2:   10 apples cost c dollars.

Estimation   8 apples would cost $2, so 10 apples will cost more than $2.

$$\frac{4 \text{ apples}}{10 \text{ apples}} = \frac{\$1}{\$c}$$

Pair 1 is the numerator of each ratio.

Pair 2 is the denominator of each ratio.

$$4c = 10 \qquad \text{Cross multiply.}$$

$$\frac{4c}{4} = \frac{10}{4} \qquad \text{Divide by the coefficient of } c.$$

$$c = 2.50 \qquad \text{To nearest cent.}$$

Interpretation   **Ten apples cost $2.50.**

(b)   Pair 1:   4 apples cost $1.

Pair 2:   a apples cost $10.

Estimation   If 10 apples cost $2.50, 4 times as many apples can be bought for $10 since $10 is 4 × $2.50. 40 apples can be bought.

$$\frac{4 \text{ apples}}{a \text{ apples}} = \frac{\$1}{\$10}$$ Pair 1.

Pair 2.

$$\$10(4) = \$1(a)$$ Cross multiply.

$$40 = a$$

Interpretation   **Forty apples can be bought for $10.**

Congruent triangles have the same size and shape. **Similar triangles** have the same shape but not the same size.

Every triangle has six parts: three angles and three sides. Each angle or side of one similar or congruent triangle has a corresponding angle or side in the other similar or congruent triangle. The symbol for showing congruency is ≅ and is read "is congruent to." The symbol for a triangle is △.

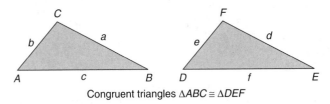

Congruent triangles △*ABC* ≅ △*DEF*

Corresponding angles of congruent triangles are the same size; that is, they are equal in measure.

Angle *A* = angle *D*

Angle *B* = angle *E*

Angle *C* = angle *F*

Corresponding sides of congruent triangles are the same size; that is, they are equal in measure.

Side *a* = side *d*

Side *b* = side *e*

Side *c* = side *f*

The symbol for showing similarity is ~ and is read "is similar to."

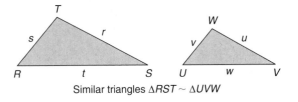

Similar triangles △*RST* ~ △*UVW*

Corresponding angles of similar triangles are equal in size.	**Corresponding sides of similar triangles** are directly proportional.
Angle R = angle U	Side r corresponds to side u
Angle S = angle V	Side s corresponds to side v
Angle T = angle W	Side t corresponds to side w

▲ **APPLICATION: Using Similar Triangles to Find Missing Dimensions**

> **How to**
>
> *Find the height of an object using the shadow and known height of another object:*
>
> 1. Note the known height of the object.
> 2. Note the length of the shadow of both the known object and the object whose height is unknown.
> 3. Let a letter equal the unknown height.
> 4. Write a proportion using the four items in Steps 1–3.
> 5. Solve the proportion.

EXAMPLE   A tree surgeon must know the height of a tree to determine which way to fell it so that it does not endanger lives, traffic, or property. A 6-ft pole casts a 4-ft shadow when the tree casts a 20-ft shadow. What is the height of the tree?

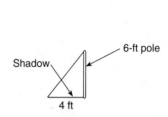
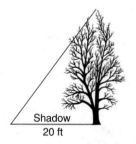

The triangles formed are similar, so we use a proportion to solve the problem. Let x = the height of the tree.

Pair 1:   6-ft pole casts a 4-ft shadow .

Pair 2: x-ft tree casts a 20-ft shadow .

Estimation   Tree is more than 6 ft tall.

$$\frac{6 \text{ (height of pole)}}{x \text{ (height of tree)}} = \frac{4 \text{ (shadow of pole)}}{20 \text{ (shadow of tree)}} \qquad \text{Pair 1.}$$
Pair 2.

$$6(20) = 4x \qquad \text{Cross multiply.}$$

$$120 = 4x \qquad \text{Divide}$$

$$30 = x$$

Interpretation   **The tree is 30 ft tall.**

▲ APPLICATION: International Currency Conversions

How to

Convert currency to a different international currency:

1. Determine the rate of exchange from one currency to the new currency.
2. Determine the amount of money to be exchanged.
3. Write a proportion that has the exchange rate as the numerators of the fractions. The amounts of money changed are the denominators. Each fraction or ratio has the same currency unit in the numerator and denominator.
4. Solve the proportion.

EXAMPLE   To change $1,000 U.S. to British pounds (£):

1.456301 U.S. dollars = 1 British pound   Determine the exchange rate.

$1,000 to be exchanged for £   Determine the amount to be exchanged.

Pair 1:
Pair 2:
$$\frac{\$ 1.456301}{\$ 1,000} = \frac{1 £}{n £}$$

Write a proportion and solve. $\frac{\$}{\$} = \frac{£}{£}$

$$1.456301\, n = 1,000(1)$$

$$n = \frac{1,000}{1.456301}$$

$$n = 686.67$$

$1,000 = 686.67 £

Linear Equations

Percentage problems can be solved using the **percentage proportion.** The percentage proportion formula is $\frac{R}{100} = \frac{P}{B}$, where R represents the **rate,** P represents the **percentage,** and B represents the **base.** The rate is divided by 100 to convert it from a percent to a fraction. The percentage proportion can be used to find the rate, percentage, or base.

How to

Solve the percentage proportion:

1. Replace appropriate variables with known facts.
2. Cross multiply to find the cross products.
3. Divide the product of the two known factors by the factor with the letter.

EXAMPLE To find 20% of 75:

The rate is 20%, the base is 75, and the percentage is missing.

$$\frac{R}{100} = \frac{P}{B}$$ Set up the proportion by replacing variables with known values.

$$\frac{20}{100} = \frac{P}{75}$$ Cross multiply.

$$20 \times 75 = 100 \times P$$

$$1500 = 100 \times P$$ Divide by coefficient of P.

$$\frac{1500}{100} = P$$ Reduce fraction.

$$15 = P$$

Therefore, the percentage is 15.

● **RELATED TOPIC:** Percentage formula

▲ **APPLICATION:** Installment Price and Monthly Payment Based on Simple Interest

How to

Find the installment price and monthly payment on an installment loan:

1. Use the percentage proportion to find the amount of interest for the loan.
2. Add the list price and interest.
3. Divide the total from Step 2 by the number of payments.

EXAMPLE   A computer that costs $1,299 is purchased on installment for 18 months, and the simple interest rate is 12%. To find the installment price and the monthly payment:

$$\frac{12}{100} = \frac{N}{\$1,299}$$ Write a proportion to find the amount of interest for 1 year.

$$100\,N = 12(\$1,299)$$

$$\frac{100N}{100} = \frac{\$15,588}{100}$$

$$N = \boxed{\$155.88}$$ Interest for 1 year.

$$\boxed{\$155.88} \times 1.5 = \boxed{\$233.82}$$ Interest for 18 months ($1\frac{1}{2}$ years).

$$\$1,299 + \boxed{\$233.82} = \$1,532.82$$ Total installment price.

$$\$1,532.82 \div 18 = \$85.16$$ Monthly payment.

How to

Find the sum of consecutive numbers beginning with 1:

1. Multiply the largest number by 1 more than the largest number.
2. Divide the product by 2.

$$\text{Sum of consecutive numbers beginning with 1} = \frac{\text{Largest number} \times (\text{Largest number} + 1)}{2}$$

Linear Equations

EXAMPLE To find the sum of the numbers from 1 through 25:

$$\frac{25(25 + 1)}{2} =$$ Multiply the largest number by 1 more than the largest number.

$$\frac{25(26)}{2} =$$

$$\frac{650}{2} =$$ Divide the product by 2.

$$325$$ Sum of the numbers from 1 through 25.

└─

▲ **APPLICATION: Interest Refund on an Early Loan Payoff**

How to

Find the interest refund using the Rule of 78:

1. Find the numerator of the refund fraction: the sum of the sequence of numbers for payments remaining.
2. Find the denominator of the refund fraction: the sum of the sequence of numbers of *all* the payments.
3. Find the interest fraction: multiply the total interest for the loan by the refund fraction.

Interest refund = Total interest × Refund fraction

If the interest on a loan or **discounted note** is paid at the beginning of the loan, a portion of the interest is refunded if the loan is paid early. To determine the amount of interest refund due, we multiply a refund fraction by the total interest. The *numerator* of the refund fraction is the sum of the period sequence numbers for the number of months remaining in the loan. The denominator of the refund fraction is the sum of the period sequence numbers for the total number of months of the loan. The **period sequence numbers** for a number of months n are the natural numbers from 1 to n. The *sum* of the period sequence numbers for a period of n months is $1 + 2 + 3 + \ldots + n$. The sum also can be found with the formula $\frac{n(n+1)}{2}$. For example, the period sequence numbers for 5 months are 1, 2, 3, 4, 5. The sum of the period sequence numbers for 5 months is $1 + 2 + 3 + 4 + 5 = 15$ or $\frac{5(5+1)}{2} = \frac{5(6)}{2} = \frac{30}{2} = 15.$

EXAMPLE   A loan for 12 months with a finance charge of $117 is paid in full with four payments remaining. To find the amount of the finance charge (interest) refund:

$$\text{Refund fraction numerator} = \frac{4(5)}{2} = \frac{20}{2} = \boxed{10}$$

Use the formula to sum a sequence of numbers.

$$\text{Refund fraction denominator} = \frac{12(13)}{2} = \frac{156}{2} = \boxed{78}$$

$$\text{Refund fraction} = \frac{\boxed{10}}{\boxed{78}}$$

Interest refund = total interest or finance charge × refund fraction

$$= \$117 \times \frac{\boxed{10}}{\boxed{78}}$$

$$= \$15$$

The finance charge refund is $15.

How to

Find the rate in an increase or decrease problem:

1. Let R represent the rate of increase or decrease.
2. Subtract the original amount from the new amount to find the amount of increase, or subtract the new amount from the original amount to find the amount of decrease.
3. Set up a proportion with R as the rate of increase or decrease, the amount of increase or decrease as the percentage (P), and the original amount as the base (B).
4. Solve the proportion to find the amount of increase or decrease.

▲ **APPLICATION: Rate of Salary Increase**

Linear
Equations

EXAMPLE   Your salary is $42,000 annually and you receive an increase to an annual salary of $43,400. To find your rate (percent) of increase:

Let R represent the rate of increase.

$43,400 - $42,000 = $1,400   Subtract the original amount from the new amount to find the amount of increase.

$$\frac{R}{100} = \frac{1,400}{42,000}$$ The proportion has 42,000 as the base and 1,400 as the percentage.

$42,000R = 1,400(100)$ Cross multiply to solve the proportion.

$42,000R = 140,000$ Divide.

$$R = \frac{140,000}{42,000}$$

$$R = 3.33\%$$

The rate of increase is 3.33%.

⌐

▲ **APPLICATION: Inverse Proportions or Variation**

An **inverse proportion** is one in which the quantities being compared are inversely related. That is, as one quantity increases, the other decreases, or as one quantity decreases, the other increases. This relationship is also called **inverse variation.**

How to

Solve an application using an inverse proportion:

1. Establish two pairs of related data.
2. Arrange one pair as the numerator of one ratio and the denominator of the other ratio.
3. Arrange the other pair so that each ratio contains like measures.
4. Form a proportion using the two ratios.
5. Solve the proportion.

EXAMPLE   A 10-in.-diameter gear is in mesh with a 5-in.-diameter gear. If the larger gear has a speed of 25 revolutions per minute (rpm), to find the rate in revolutions per minute (rpm) of the smaller gear:

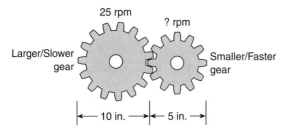

Because gears in mesh are *inversely* related, we set up an inverse proportion. Each ratio uses like measures and the ratios are in inverse order.

Pair 1:   25 rpm for 10-in. size of larger gear.

Pair 2:   *x* rpm or 5-in. size of smaller gear.

Estimation We expect the speed of the smaller gear to be faster than the 25-rpm speed of the larger gear. The speed of the smaller gear will be more than 25 rpm.

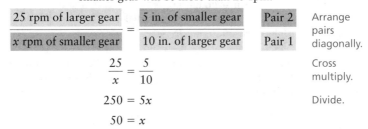

$$\frac{25 \text{ rpm of larger gear}}{x \text{ rpm of smaller gear}} = \frac{5 \text{ in. of smaller gear}}{10 \text{ in. of larger gear}} \quad \begin{matrix}\text{Pair 2} \\ \text{Pair 1}\end{matrix} \quad \begin{matrix}\text{Arrange}\\ \text{pairs}\\ \text{diagonally.}\end{matrix}$$

$$\frac{25}{x} = \frac{5}{10} \qquad \text{Cross multiply.}$$

$$250 = 5x \qquad \text{Divide.}$$

$$50 = x$$

Interpretation **The smaller gear turns at the faster speed of 50 rpm.**

How to distinguish between direct and inverse variation

We can distinguish between direct and inverse variation by anticipating cause-and-effect situations.

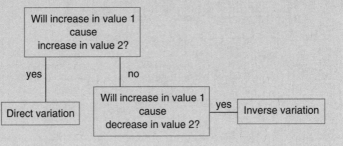

Linear Equations

■ 7–4 Equations Containing One Absolute-Value Term

Equations containing absolute-value terms are not linear equations. These equations may have as many as two solutions. To solve an equation with one or more absolute value terms, we use the basic concepts for solving linear equations.

How to

Solve an equation containing one absolute-value term:

1. Isolate the absolute-value term on one side of the equation.
2. Determine if the equation has no, one, or two solutions.
   a. $|x| = b$ has no solution if b is negative ($b < 0$).
   b. $|x| = b$ has one solution if $b = 0$.
   c. $|x| = b$ has two solutions if b is positive ($b > 0$).
3. If the equation has two solutions, separate the equation into two cases. One case considers the expression within the absolute-value symbol to be positive $x = b$. The other case considers it to be negative $-x = b$.
4. Solve each case to obtain the *two* roots of the equation.
5. Check each solution.

EXAMPLE   To find the roots of the equation $|y + 3| - 5 = 6$:
Before we can separate the equation into two cases, we *isolate* the absolute-value term; that is, we rearrange the terms of the equation so that the absolute-value term is alone on one side of the equation.

$$|y + 3| - 5 = 6 \qquad \text{Isolate the absolute value term.}$$

$$|y + 3| = 6 + 5 \qquad \text{Combine like terms.}$$

$$|y + 3| = 11 \qquad \text{Equation will have two solutions. Separate into two cases.}$$

Case 1: $y + 3 = 11$	*Case 2:* $-(y + 3) = 11$
$y = 11 - 3$	$-y - 3 = 11$

$$y = 8$$

$$-y = 11 + 3$$
$$-y = 14$$
$$y = -14$$

The roots of the equation are 8 and -14.

Check:

Case 1: $|y + 3| = 11$

$$|\,8\, + 3| = 11$$
$$|11| = 11$$
$$11 = 11$$

Case 2: $|y + 3| = 11$

$$|-14 + 3| = 11$$
$$|-11| = 11$$
$$11 = 11$$

● **RELATED TOPICS: Absolute value, Linear equations**

■ 7–5 Formula Evaluation and Rearrangement

How to

Evaluate a formula:

1. Write the formula.
2. Rewrite the formula substituting known values for variables of the formula.
3. If the unknown variable is isolated, use the order of operations to evaluate. If the unknown variable is *not* isolated, solve the equation from Step 2 for the missing variable.
4. Interpret the solution within the context of the formula.

EXAMPLE   To solve the formula $P = 2(l + w)$ for w if $P = 12$ ft and $l = 4$ ft:

$P = 2(l + w\,)$ Perimeter of a rectangle = two × the sum of the length and the width.

$12 = 2(4 + w\,)$ Substitute values. Apply the distributive property.

$12 = 8 + 2\,w$ Isolate the term with the variable.

Linear
Equations

$$12 - 8 = 2\,w \qquad \text{Combine like terms.}$$

$$4 = 2\,w \qquad \text{Divide by the coefficient of } w.$$

$$\frac{4}{2} = \frac{2\,w}{2}$$

$$2 = w \qquad \text{Interpret solution.}$$

The width is 2 ft.

⎿

● **RELATED TOPICS: Distributive property, Perimeter of a rectangle**

How to

Rearrange a formula:

1. Determine which variable of the formula will be isolated (solved for).
2. Highlight or mentally locate all instances of the variable to be isolated.
3. Treat all other variables of the formula as you would numbers in an equation and perform normal techniques for solving an equation.
4. If the isolated variable is on the right side of the equation, interchange the sides so that the variable appears on the left side.

EXAMPLE   Solve for R in the formula $I = PRT$.

$$I = P\,R\,T \qquad \begin{array}{l} \text{Simple interest} = \text{principle} \times \text{rate} \times \text{time} \\ \text{We are solving for } R, \text{ therefore, } PT \text{ is its coefficient.} \end{array}$$

$$\frac{I}{PT} = \frac{P\,R\,T}{PT} \qquad \text{Divide both sides by the coefficient of the variable } R.$$

$$\frac{I}{PT} = R \qquad \text{Rewrite with } R \text{ on the left.}$$

$$R = \frac{I}{PT}$$

⎿

● **RELATED TOPIC: Simple interest**

CHAPTER **8**

Graphing Linear Equations

- 8–1 Linear Equations in Two Variables and Function Notation
- 8–2 Graphing Linear Equations and Functions with Two Variables

8–1 Linear Equations in Two Variables and Function Notation

The number line shows values visually and is called a **one-dimensional graph**. It takes one number to locate a point on a one-dimensional graph. The **rectangular coordinate system** gives a graphical representation of two-dimensional values. It takes two numbers to locate a point on a two-dimensional graph. In the rectangular coordinate system, two number lines are positioned to form a right angle or square corner.

The number line that runs from left to right is the **horizontal axis** and is represented by the letter x. The number line that runs from top to bottom is the **vertical axis** and is represented by the letter y.

● RELATED TOPIC: Number line

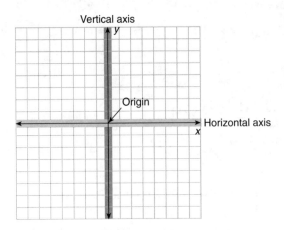

Vertical axis

y

Origin

Horizontal axis

x

Zero on both number lines is located at the point where the two number lines cross. This point is called the **origin**. Other points are located by horizontal and vertical movements from the origin.

Symbolically, we represent the location of a point as

(horizontal movement, vertical movement)   or (x, y)

where x is the horizontal movement from the origin, or the x-coordinate, and y is the vertical movement from the x-axis, or the y-coordinate. The notation (x, y) is called **point notation**. The number pair in point notation is called an **ordered pair** since the numbers *must* be written in a specific order, the horizontal value first.

How to

Plot a point on the rectangular coordinate system:

1. Start at the origin.
2. Count to the left or right the number of units indicated by the first signed number (x-coordinate).
3. From the ending point from Step 2, count up or down the number of units indicated by the second signed number (y-coordinate).

EXAMPLE To plot the points: Point $A = (3, 1)$, point $B = (-2, 5)$, point $C = (-3, -2)$, point $D = (1, -3)$, point $E = (0, -2)$, point $F = (5, 0)$: Start at the origin for each point.

Point A (3, 1):   right 3, up 1

Point B (−2, 5):   left 2, up 5

Point C (−3, −2):   left 3, down 2

Point D (1, −3):   right 1, down 3

Point E (0, −2):   no horizontal movement, down 2

Point F (5, 0):   right 5, no movement vertically

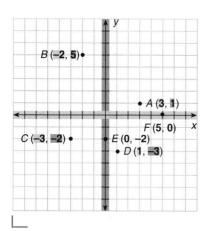

A **linear equation** in two variables is an equation that has a straight line as its graph. It can be written in the **standard form**

$$ax + by = c$$

where *a, b,* and *c* are real numbers and *a* and *b* are not both zero.

If either *a* or *b* is zero, the equation is a linear equation in *one* variable. In Chapter 7, linear equations in one variable were solved algebraically.

The solutions of a linear equation in two variables are **ordered pairs** of numbers. There is one number in the ordered pair for each variable. The numbers are generally ordered in the alphabetical order of the two variables rather than in the order the letters appear in the equation. Thus, in the equation $y = -2x + 5$ a solution of (1, 3) indicates that $x = 1$ and $y = 3$. An ordered pair is written in point notation (enclosed in parentheses and the numbers separated by a comma).

Linear equations in two variables are commonly written in one of three forms.

Standard Form	Solved for *y*	Function Notation
$2x + y = 5$	$y = -2x + 5$	$f(x) = -2x + 5$

The **standard form** of a linear equation in two variables is written with both variables on the left side of the equation and the independent variable written first with a positive coefficient. The number or constant is written on the right side of the equation.

When the equation is **solved for y**, the right side of the equation has the independent variable term followed by the constant. This form is referred to as the **slope-intercept form of a linear equation.**

Function notation is a notation showing the relationship between an **independent variable** (also called **input variable**) and a **dependent variable** (also called **output variable**) by using the symbol $f(x)$. This symbol is read "a function of x" or "f of x." It does *not* mean "f times x." The symbol $f(x)$ represents the **dependent variable** in function notation just as y represents the dependent variable in standard form and the slope-intercept form.

In each form x is the **independent variable**. The value of the independent variable, x, is selected from a set of numbers called the **domain**. The values in the domain are sometimes referred to as **input values**. The value of $f(x)$ is the result of evaluating the equation or function for a particular value of x. All possible values of the dependent variable make up a set called the **range**. The values from the range are sometimes referred to as **output values.**

There are many ordered pairs that solve a linear equation in two variables. Some of these solutions can be written in a **table of values** or **table of solutions.**

How to

Prepare a table of solutions for an equation or function with two variables:

1. Select any value for the independent variable.
2. Substitute the selected value for the independent variable and solve the resulting equation in one variable.
3. Repeat Steps 1 and 2 until the desired number of solutions are obtained.

EXAMPLE   To make a table of solutions for the equation $y = -2x + 5$: Select at least three values for the independent variable: $-1, 0, 1$.

when $x = -1, y = -2(-1) + 5$ Substitute -1 for x.

$\qquad\qquad = 2 + 5$ Evaluate.

$\qquad\qquad = 7$ Solution: $x = -1$ and $y = 7$, or $(-1, 7)$.

when $x = 0$, $y = -2(0) + 5$ Substitute.

$\qquad\qquad = 0 + 5$ Evaluate.

$\qquad\qquad = 5$ Solution: (0, 5).

when $x = 1$, $y = -2(1) + 5$ Substitute.

$\qquad\qquad = -2 + 5$ Evaluate.

$\qquad\qquad = 3$ Solution: (1, 3).

Make a table of the three solutions.

x	y
−1	7
0	5
1	3

● **RELATED TOPIC:** Evaluating formulas or equations

■ 8–2 Graphing Linear Equations and Functions with Two Variables

How to

Graph a linear equation or function using a table of values:

1. Prepare a table of values or solutions by evaluating the function at different values (at least three) of the independent variable.
2. Plot the points in the table of values on the rectangular coordinate system, with the independent variable on the x-axis and the dependent variable on the y-axis.
3. Connect the points with a straight line. Extend the graph beyond the three points and place an arrow on each end to indicate that the line extends indefinitely in both directions.

● **RELATED TOPICS:** Graph of nonlinear equations, Graphs of quadratic equations, Plotting points on a rectangular coordinate system

EXAMPLE   To graph the function $y = -2x + 5$:
Plot the solutions in the table of solutions as points on a rectangular coordinate system.

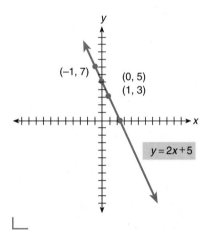

⌐

▲ APPLICATION: Cost Function

A **cost function** is an equation that shows the cost of an item or a service based on a fixed cost and a variable cost.

The **fixed cost** is a constant that represents a portion of the cost no matter how many items are purchased or what amount of service is used. The **variable cost** represents the portion of the cost that is based on the number of items purchased or the amount of service used.

EXAMPLE   A cellular phone monthly service contract has a fixed cost of $15 and a variable cost of $0.25 per minute for each minute used in the month. To make a table of values and graph the cost function $f(x) = 15 + 0.25x$ for five values of x in 50-min intervals:

$f(0) = 15 + 0.25(0)$ $f(50) = 15 + 0.25(50)$ $f(100) = 15 + 0.25(100)$

$f(0) = 15 + 0$ $f(50) = 15 + 12.5$ $f(100) = 15 + 25$

$f(0) = 15$ or 15.00 $f(50) = 27.5$ or 27.50 $f(100) = 40$ or 40.00

$f(150) = 15 + 0.25(150)$ $f(200) = 15 + 0.25(200)$

$f(150) = 15 + 37.5$ $f(200) = 15 + 50$

$f(150) = 52.5$ or 52.50 $f(200) = 65$ or 65.00

x	f(x)
0	$15.00
50	$27.50
100	$40.00
150	$52.50
200	$65.00

To represent this model graphically, use the *x*- or horizontal axis to represent the values of the independent variable. Use the *y*- or vertical axis to represent the values of the dependent variable.

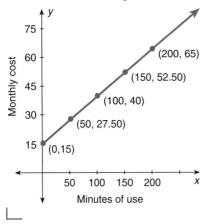

Data like that shown in the table of solutions and the graph in the previous example are referred to as **paired data**.

● **RELATED TOPICS: Graphing linear equations, Statistics**

The **x-intercept** is the point on the *x*-axis through which the line of the equation passes; that is, the *y*-coordinate is zero $(x, 0)$.

The **y-intercept** is the point on the *y*-axis through which the line of the equation passes; that is, the *x*-coordinate is zero $(0, y)$.

How to

Find the intercepts of a linear equation:

1. To find the *x*-intercept, let $y = 0$ and solve for *x*.
2. To find the *y*-intercept, let $x = 0$ and solve for *y*.

EXAMPLE   To find the intercepts of the equation $3x - y = 5$:

$3x - y = 5$ For the x-intercept, let $y = 0$.

$3x - 0 = 5$ Solve for x.

$3x = 5$

$x = \dfrac{5}{3}$ Write as a mixed number or decimal equivalent.

$x = 1\dfrac{2}{3}$ or 1.67 A mixed number or decimal is easier to plot than an improper fraction.

The x-intercept is $(1\frac{2}{3}, 0)$.

$3x - y = 5$ For the y-intercept, let $x = 0$.

$3(0) - y = 5$ Solve for y.

$-y = 5$

$y = -5$

The y-intercept is $(0, -5)$.

How to

Graph a linear equation by the intercepts method:

1. Find the x- and y-intercepts.
2. Plot the intercepts on a rectangular coordinate system.
3. Draw the line through the two points and extend it beyond each point.
4. Check by examining one additional solution of the equation.

EXAMPLE   To graph the equation $3x - y = 5$ by using the intercepts of each axis:

Plot the two intercepts found in the preceding example, $(1\frac{2}{3}, 0)$ and $(0, -5)$. Draw the line connecting these points and extending beyond each point. Check by finding one other point and plotting it. If it is on the line, the graph is correct. To find a third point, let $x = 2$:

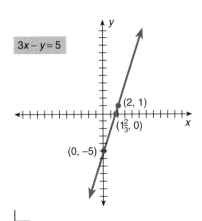

$$3x - y = 5$$ Substitute 2 for x.

$$3(2) - y = 5$$ Solve for y.

$$6 - 5 = y$$

$$1 = y$$ The check point (2, 1) is on the graph.

TIP

Graphing equations when both intercepts are (0, 0)

If both intercepts are (0, 0), they coincide on the origin, and form only *one point*. An additional point must be found by the table-of-values method so that you will have two distinct points for drawing the line. A third point is still useful to check your work.

$$y = 7x$$

If $x = 0$, then $y = 7(0)$ or $y = 0$. Both the x- and y-intercepts are (0, 0). So let $x = 1$, then $y = 7(1)$ or $y = 7$. *Plot* the points (1, 7) and (0, 0) and draw the graph.

EXAMPLE   To graph $y = 7x$ using the intercepts method:

Plot the points (1, 7) and (0, 0) as shown in the Tip feature. Draw the line through the two points and extend it beyond each point, as shown in the figure.

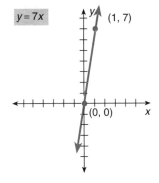

The **slope** of a line is the ratio of the vertical change to the horizontal change.

A slope of 1 (written as $\frac{1}{1}$) means a change of 1 vertical unit for every 1 horizontal unit of change. A slope of 3 (written as the ratio $\frac{3}{1}$) means a change of 3 vertical units for every 1 horizontal unit. A slope of $\frac{1}{3}$ means a change of 1 vertical unit for every 3 horizontal units. This slope can be determined from any two points on the graph.

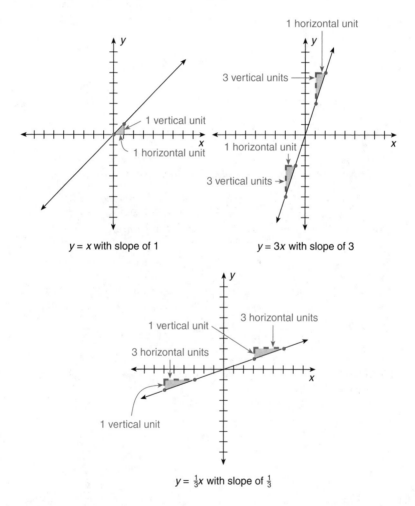

$y = x$ with slope of 1 $y = 3x$ with slope of 3

$y = \frac{1}{3}x$ with slope of $\frac{1}{3}$

The **slope-intercept form of a linear equation** is $y = mx + b$, where m is the slope and b is the y-coordinate of the y-intercept. When equations are written in this form, the slope and y-intercept can be identified by *inspection*.

How to

Find the slope and y-intercept of a linear equation in two variables:

1. Solve the equation for y.
2. The slope is the coefficient of x.
3. The y-coordinate of the y-intercept is the constant.

Symbolically, in the equation of $y = mx + b$,

slope $= m$ y-intercept $= (0, b)$

EXAMPLE   To write the equations in slope-intercept form and identify the slope and y-intercept:

(a) $2x + y = 4$ (b) $5x - y = -2$ (c) $3x + 4y = -12$

(a) $2x + y = 4$	Solve for y. Add $-2x$ to both sides.
$y = -2x + 4$	
slope $= -2$ or $\dfrac{-2}{1}$ or $\dfrac{2}{-1}$	The slope is the coefficient of x.
y-intercept $= 4$ or $(0, 4)$	The y-intercept is the constant.
(b) $5x - y = -2$	Solve for y. Add $-5x$ to both sides.
$\dfrac{-y}{-1} = \dfrac{-5x - 2}{-1}$	Divide by -1.
$y = 5x + 2$	
slope $= 5$ or $\dfrac{5}{1}$	The slope is the coefficient of x.
y-intercept $= 2$ or $(0, 2)$	The y-intercept is the constant.
(c) $3x + 4y = -12$	Solve for y. Sort terms to isolate $4y$. Divide by 4.
$\dfrac{4y}{4} = \dfrac{-3x - 12}{4}$	Write the right side as separate terms and reduce.

$$y = -\frac{3}{4}x - 3$$

slope $= -\dfrac{3}{4}$ or $\dfrac{-3}{4}$ or $\dfrac{3}{-4}$ The slope is the coefficient of x.

y-intercept $= -3$ or $(0, -3)$ The y-intercept is the constant.

⌞

● **RELATED TOPICS: Dividing polynomials, Rearranging formulas**

An equation in the slope-intercept form can be graphed using only the slope and y-intercept.

How to

Graph a linear equation in the form $y = mx + b$ using the slope and y-intercept:

1. From the equation identify the y-intercept and locate the y-intercept on the y-axis.
2. Using the slope, determine the amount of vertical and horizontal movement indicated.
3. From the y-intercept, locate additional points on the graph of the equation by counting the indicated vertical and horizontal movement.
4. Draw the line connecting the points, and extend it beyond the points.

EXAMPLE   To graph the equation $2x + y = 4$ using the slope and y-intercept:

$2x + y = 4$ Write the equation in slope-intercept form.

$y = -2x + 4$ Solve for y. The y-intercept is 4, or (0, 4) in point notation.

y-intercept $= (0, 4)$ Locate this point on the y-axis.

slope $= -2$ or $\dfrac{-2}{1}$ or $\dfrac{2}{-1}$ Write the slope, -2, in fraction form. Use the slope to determine the vertical and horizontal movement between two points.

$\frac{-2}{1}$ indicates vertical movement of -2 and horizontal movement of $+1$ from the y-intercept. $\frac{2}{-1}$ indicates vertical movement of $+2$ and horizontal movement of -1.

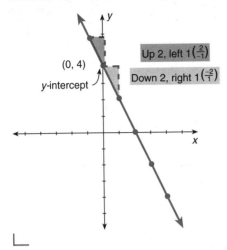

(0, 4)

y-intercept

Up 2, left 1 $\left(\frac{2}{-1}\right)$

Down 2, right 1 $\left(\frac{-2}{1}\right)$

CHAPTER 9

Slope and Distance

- **9–1 Slope**
- **9–2 Point–Slope Form of an Equation**
- **9–3 Slope–Intercept Form of an Equation**
- **9–4 Parallel and Perpendicular Lines**
- **9–5 Distance and Midpoints**

■ 9–1 Slope

The **slope** of a line is the ratio of the vertical rise of
the line to the horizontal run of the line.

$$\text{Slope} = \frac{\text{Rise}}{\text{Run}}$$

Rise

Run

How to

Find the slope of a line from two given points on the line:

1. Designate one point as point 1 with coordinates (x_1, y_1). Designate
   the other point as point 2 with coordinates (x_2, y_2).
2. Calculate the change (*difference*) in the y-coordinates to find the
   vertical *rise* $(y_2 - y_1)$ and the x-coordinates to find the horizontal
   run $(x_2 - x_1)$.
3. Write a ratio of the rise to the run and reduce the ratio to lowest terms.

Symbolically, if P_1 and P_2 are any two points on the line,

$$\text{Slope} = \frac{\Delta y}{\Delta x} = \frac{y_2 - y_1}{x_2 - x_1}$$

where $P_1 = (x_1, y_1)$ and $P_2 = (x_2, y_2)$

The Greek capital letter delta (Δ) is used to indicate a change and
to write the slope definition symbolically.

EXAMPLE   To find the slope of a line if the points $(2, -1)$ and $(5, 3)$ are on the line:

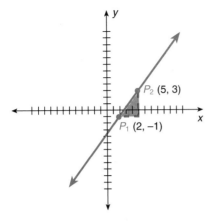

Label $(2, -1)$ as point 1 (P_1) and $(5, 3)$ as point 2 (P_2).

Change in $y = \Delta y = \text{rise} = y_2 - y_1 = 3 - (-1) = 4$

Change in $x = \Delta x = \text{run} = x_2 - x_1 = 5 - 2 = 3$

$$\text{Slope} = \frac{\Delta y}{\Delta x} = \frac{\text{rise}}{\text{run}} = \frac{y_2 - y_1}{x_2 - x_1} = \frac{3 - (-1)}{5 - 2} = \frac{4}{3}$$

The slope of the line through points $(2, -1)$ and $(5, 3)$ is $\frac{4}{3}$.

⌐

TIP

What points on a line are used? In what order are they used?

Any line has an infinite number of points. Any two points on the line can be used to find the slope. Also, the designation of P_1 and P_2 is not critical. Let's find the slope of the line in the preceding example by designating P_1 as $(5, 3)$ and P_2 as $(2, -1)$.

$\Delta y = \text{Rise} = y_2 - y_1 = -1 - 3 = -4$

$\Delta y = \text{Run} = x_2 - x_1 = 2 - 5 = -3$

$\dfrac{\Delta y}{\Delta x} = \dfrac{\text{Rise}}{\text{Run}} = \dfrac{-4}{-3} = \dfrac{4}{3}$

When the y-coordinates of two points are the same, the line through these two points is a horizontal line. Any horizontal line has no change in y-coordinates, or no rise. **The slope of any horizontal line is zero.**

EXAMPLE   To find the slope of the horizontal line that passes through $(3, 2)$ and $(-3, 2)$:
Let $P_1 = (3, 2)$ and $P_2 = (-3, 2)$.

$$\text{Slope} = \frac{\Delta y}{\Delta x} = \frac{y_2 - y_1}{x_2 - x_1} = \frac{2 - 2}{-3 - 3} = \frac{0}{-6} = 0 \qquad \text{The } y\text{-coordinates are the same.}$$

When the x-coordinates of two points are the same, the line through these two points is a vertical line. Any vertical line has no change in x-coordinates or no run. **The slope of any vertical line is undefined.** We say that a vertical line has *no slope*.

EXAMPLE   To find the slope of the vertical line that passes through the points $(3, 2)$ and $(3, -2)$:
Let $P_1 = (3, 2)$ and $P_2 = (3, -2)$.

$$\text{Slope} = \frac{\Delta y}{\Delta x} = \frac{y_2 - y_1}{x_2 - x_1} = \frac{-2 - 2}{3 - 3} = \frac{-4}{0} \qquad \text{The } x\text{-coordinates are the same.}$$

Recall that division by zero is impossible; **therefore, the slope of the vertical line passing through $(3, 2)$ and $(3, -2)$ is undefined.**

TIP

Slope of zero versus no slope

A slope of zero and no slope are *not* the same. Zero is a real number. A slope of zero is real. When zero is the denominator of a fraction, the number is undefined. "No slope" is the same as an undefined slope.

How to

Identify a horizontal or vertical line:

1. Examine the y-coordinates of two points on the line. If the y-coordinates are the same, the line is horizontal and the slope is 0.
2. Examine the x-coordinates of two points on the line. If the x-coordinates are the same, the line is vertical and the slope is undefined.

EXAMPLE   To identify the line passing through each pair of points as horizontal, vertical, or neither.

(a) $(7, 2)$ and $(-3, 2)$ (b) $(-5, 2)$ and $(-5, -4)$

(c) $(2, -2)$ and $(5, -5)$

(a) $(7, 2)$ and $(-3, 2)$ The y-values are the same.

The line is horizontal.

(b) $(-5, 2)$ and $(-5, 4)$ The x-values are the same.

The line is vertical.

(c) $(2, -2)$ and $(5, -5)$ Neither the y-values or the x-values are the same.

The line is neither horizontal nor vertical.

■ 9–2 Point–Slope Form of an Equation

The equation of a line can be determined from the slope and one point on the line.

How to

Find the equation of a line if the slope and one point on the line are known:

1. Use either version of the **point-slope form of an equation** of a straight line.

$$y - y_1 = m(x - x_1) \quad \text{or} \quad m = \frac{y - y_1}{x - x_1}$$

where x_1 and y_1 are coordinates of the known point, m is the slope of the line, and x and y are the variables of the equation.
2. Substitute known values for x_1, y_1 and m.
3. Rearrange the equation to be in standard form ($ax + by = c$) or solved for y ($y = mx + b$).

EXAMPLE   To find the equation of the line with a slope of $\frac{2}{3}$ and passing through the point $(3, -2)$:

$y - y_1 = m(x - x_1)$ Substitute into the point-slope form.

$y - (-2) = \dfrac{2}{3}(x - 3)$ $m = \dfrac{2}{3}$; $x_1 = 3$; $y_1 = -2$

$$y + 2 = \frac{2}{3}(x - 3) \qquad \text{Distribute. } \frac{2}{\underset{1}{3}} \cdot \frac{-\overset{-1}{3}}{1} = -2$$

$$y + 2 = \frac{2}{3}x - 2 \qquad \text{Sort terms.}$$

$$y = \frac{2}{3}x - 2 - 2 \qquad \text{Combine like terms.}$$

$$y = \frac{2}{3}x - 4 \qquad \text{Solved for } y.$$

TIP

Standard Form of an Equation

The standard form of a linear equation is

$$ax + by = c$$

The characteristics of an equation in standard form are:

- All variable terms are on the left.
- The leading term has the x variable and is positive.
- The equation contains no fractions.

EXAMPLE   To write $y = \frac{2}{3}x - 4$ in standard form:

$$y = \frac{2}{3}x - 4 \qquad \text{Rearrange with variable terms on left with the } x\text{-variable as the first term.}$$

$$-\frac{2}{3}x + y = 4 \qquad \text{Clear fraction by multiplying by denominator 3.}$$

$$3\left(-\frac{2}{3} + y\right) = 3(4) \qquad \text{Distribute.}$$

$$-2x + 3y = 12 \qquad \text{Multiply by } -1 \text{ to make the leading term positive.}$$

$$2x - 3y = -12 \qquad \text{Standard form.}$$

How to

Find the equation of a line if two points on the line are known:

1. Use the slope formula to find the slope given two points.

$$m = \frac{y_2 - y_1}{x_2 - x_1}$$

2. Use the point-slope form of an equation of a straight line, the calculated slope from Step 1, and the coordinates of one of the given points.

$$(y - y_1) = m(x - x_1)$$

● **RELATED TOPICS: Point-slope form of an equation, Slope**

EXAMPLE   To find the equation of the line that passes through the points $(0, 8)$ and $(5, 0)$:

Find the slope of the line passing through the points $(0, 8)$ and $(5, 0)$. Let $P_1 = (\,0,\,8\,)$ and $P_2 = (\,5,\,0\,)$.

$$\text{Slope} = \frac{\Delta y}{\Delta x} = \frac{y_2 - y_1}{x_2 - x_1} = \frac{0 - 8}{5 - 0} = -\frac{8}{5} \qquad \text{Slope formula.}$$

Now use the slope and one of the points and the point-slope form of an equation to write the desired equation. Using P_1 we have

$$y - y_1 = m(x - x_1) \qquad \text{Point-slope form. Substitute values. } m = -\frac{8}{5};$$
$$\qquad\qquad\qquad\qquad x_1 = 0;\, y_1 = 8$$

$$y - 8 = -\frac{8}{5}(x - 0) \qquad \text{Distribute.}$$

$$y - 8 = -\frac{8}{5}x \qquad \text{Solve for } y.$$

$$y = -\frac{8}{5}x + 8$$

Suppose we use P_2 $(5, 0)$ instead of P_1 in the point-slope form of the equation. Because only one line passes through $(0, 8)$ and $(5, 0)$, we have the same equation if either point is used. Let's find the equation of the line passing through $(0, 8)$ and $(5, 0)$ using the point $(5, 0)$ and the slope $-\frac{8}{5}$.

$$y - y_1 = m(x - x_1) \qquad\qquad m = -\frac{8}{5};\, x_1 = 5;\, y_1 = 0$$

$$y - 0 = -\frac{8}{5}(x - 5) \qquad \text{Distribute.}$$

$$y = -\frac{8}{5}x - \frac{8}{5}(-5)$$

$$y = -\frac{8}{5}x + 8 \qquad \text{Same equation as before.}$$

A horizontal line has a slope of zero. The **equation of a horizontal line** becomes

$$y - y_1 = m(x - x_1)$$
$$y - y_1 = 0(x - x_1)$$
$$y - y_1 = 0$$
$$y = y_1 \qquad y_1 \text{ is the } y\text{-coordinate of any point on the line.}$$

Equations of horizontal and vertical lines

The equations of horizontal and vertical lines are special cases.

Horizontal line: $y = k$ where k is the common y-coordinate for all points on the line.

Vertical line: $x = h$ where h is the common x-coordinate for all points on the line.

■ 9–3 Slope–Intercept Form of an Equation

How to

Find the equation of a line if the slope and y-intercept are known:

1. Use the slope-intercept form of an equation of a straight line:

$$y = mx + b$$

where m = slope and b = the y-coordinate of the y-intercept.
2. Substitute known values for m and b.
3. Write the equation in standard form or slope–intercept form.

EXAMPLE   To write the equation for a line with a slope of -3 and a y-intercept of 5:

Slope $= m = -3;$ y-intercept $= b = 5$

$y = mx + b$ Substitute values.

$y = -3x + 5$

■ 9–4 Parallel and Perpendicular Lines

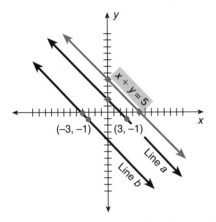

Parallel lines are two or more lines that are the same distance apart everywhere. They have no points in common. **The slopes of parallel lines are equal,** but the x- and y-intercepts are different.

> **How to**
>
> *Find the equation of a line that is parallel to a given line when at least one point on the parallel line is known:*
>
> 1. Solve the equation of the given line for y.
> 2. Determine the slope m from $y = mx + b$.
> 3. The slope of the parallel line is the same as the slope of the given line.
> 4. Use the point-slope form of a straight line $y - y_1 = m(x - x_1)$ and substitute values for m, x_1, and y_1.
> 5. Rearrange the equation to be in standard form ($ax + by = c$) or solved for y ($y = mx + b$).

EXAMPLE   To find the equation of a line that is parallel to $2y = 3x + 8$ and passes through the point $(2, 1)$:
Find the slope of the given line, $2y = 3x + 8$.

$2y = 3x + 8$ Solve for y.

$$\frac{2y}{2} = \frac{3x + 8}{2}$$

$y = \dfrac{3}{2}x + 4$ Identify slope.

The slope of $2y = 3x + 8$ is $\dfrac{3}{2}$.

Find the equation of a line passing through the point $(2, 1)$ that is parallel to $2y = 3x + 8$.

$y - y_1 = m(x - x_1)$ Substitute. The slope of a parallel line is equal to the slope of the given line.

$y - 1 = \dfrac{3}{2}(x - 2)$ $m = \dfrac{3}{2}$; $x_1 = 2$; $y_1 = 1$. Distribute.

$y - 1 = \dfrac{3}{2}x - 3$ Solve for y.

$y = \dfrac{3}{2}x - 3 + 1$ Combine like terms.

$y = \dfrac{3}{2}x - 2$ Slope-intercept form.

⬤ **RELATED TOPICS: Point-slope form, Slope intercept form**

Perpendicular lines are two lines that intersect to form right angles (90° angles). Another term for perpendicular is **normal**. The slope of *any* line

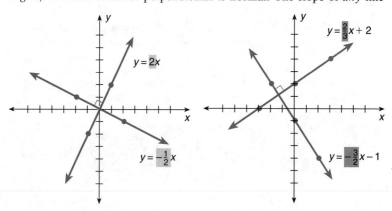

perpendicular to a given line is the **negative reciprocal** of the slope of the given line.

TIP

Negative reciprocals

The negative reciprocal of a number is not necessarily a negative value. It is the opposite of the given number. To find the negative reciprocal:

1. Interchange the numerator and denominator.
2. Take the opposite of the reciprocal.

The negative reciprocal of -5 is $+\frac{1}{5}$. The negative reciprocal of $-\frac{3}{4}$ is $+\frac{4}{3}$. The negative reciprocal of $\frac{4}{5}$ is $-\frac{5}{4}$. The negative reciprocal of 3 is $-\frac{1}{3}$.

How to

Find the equation of a line that is perpendicular (normal) to a given line and passes through a given point on the line:

1. Determine the slope of the *given* line.
2. Find the *negative reciprocal* of this slope. The negative reciprocal is the slope of the perpendicular line.
3. Write the equation for the perpendicular line by substituting the coordinates of the *given* point for x_1 and y_1 and the slope of the *perpendicular* line for m into the point-slope form of the equation $y - y_1 = m(x - x_1)$.
4. Write the equation in slope-intercept or standard form.

EXAMPLE   To find the equation of the line perpendicular to $4x + y = -3$ and passing through $(0, -3)$:

$$4x + y = -3 \qquad \text{Solve the given equation for } y.$$

$$y = -4x - 3 \qquad \text{Identify the slope.}$$

$$\text{Slope}_{given} = -4$$

$$\text{Slope}_{perpendicular} = +\frac{1}{4} \qquad \text{Negative reciprocal of } -4.$$

$$y - y_1 = m(x - x_1) \qquad \text{Point-slope form. Substitute.}$$

$$y - (-3) = \frac{1}{4}(x - 0) \qquad m = \frac{1}{4}, \, P_1 = (0, -3)$$

Slope and Distance

$$y + 3 = \frac{1}{4}x \qquad \text{Solve for } y.$$

$$y = \frac{1}{4}x - 3 \qquad \text{Slope-intercept form.}$$

or

$$4y = x - 12 \qquad \text{Multiply by } x \text{ to clear fractions and rearrange.}$$

$$x - 4y = 12 \qquad \text{Standard form.}$$

TIP

Slope-intercept form versus standard form

The variations of the equation in the preceding example represent the same line. The slope-intercept form is useful for determining properties of a graph by inspection. The standard form is desirable because it contains no fractions.

■ 9–5 Distance and Midpoints

The distance between two points on a line is the difference between the coordinates of the points on the line.

How to

Find the distance between two points on a line:

1. Determine the coordinates for each point.
2. Subtract the value of the coordinate of the leftmost point from the value of the coordinate of the rightmost point.

EXAMPLE To find the distance from $2\frac{1}{2}$ to $4\frac{1}{4}$:
Visualize the points on a number line.

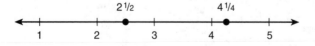

$4\frac{1}{4} - 2\frac{1}{2} =$ Subtract the coordinate of the leftmost point from the coordinate of the rightmost point.

$4\frac{1}{4} - 2\frac{2}{4} =$

$3\frac{5}{4} - 2\frac{2}{4} =$

$1\frac{3}{4}$ Distance from $2\frac{1}{2}$ to $4\frac{1}{4}$.

└

● **RELATED TOPIC: Subtracting fractions**

The distance between two points on the rectangular coordinate system is the shortest distance between the points.

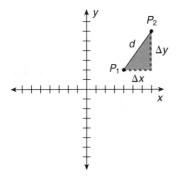

| **How to** |

Find the distance between two points on the rectangular coordinate system:

1. Use the formula

$$d = \sqrt{(x_2 - x_1)^2 + (y_2 - y_1)^2}$$

where P_1 and P_2 are two points with coordinates (x_1, y_1) and (x_2, y_2), respectively.
2. Substitute values for x_1, x_2, y_1, and y_2.
3. Evaluate to find d.

EXAMPLE To find the distance from (4, 2) to (7, 6):
Point (4, 2) is labeled P_1, and point (7, 6) is labeled P_2.

$d = \sqrt{(x_2 - x_1)^2 + (y_2 - y_1)^2}$ Substitute: $x_1 = 4$, $x_2 = 7$, $y_1 = 2$, and $y_2 = 6$.

$d = \sqrt{(7 - 4)^2 + (6 - 2)^2}$ Simplify each grouping.

$d = \sqrt{3^2 + 4^2}$ Square each term in the radicand.

$d = \sqrt{9 + 16}$ Add terms in the radicand.

$d = \sqrt{25}$ Find the principal square root.

$d = 5$

The distance from (4, 2) to (7, 6) is 5 units.

● **RELATED TOPIC: Pythagorean theorem**

Two points determine a **line** that is infinite in length. However, the two points also determine a **line segment** that has a certain length. The two points are called the **end points** of the line segment.

On a number line or measuring device, the *coordinate of the midpoint between two points* is the average of the coordinates of the points.

How to

Find the midpoint of a line segment formed by two points on a line:

1. Use the formula

$$\text{Midpoint} = \frac{P_1 + P_2}{2}$$

where P_1 and P_2 are points on the number line or measuring device.

2. Substitute values for P_1 and P_2.
3. Evaluate to find the midpoint.

EXAMPLE   To find the midpoint between two points on a metric rule at 2.8 and 5.6:

$$\text{Midpoint} = \frac{P_1 + P_2}{2} \qquad \text{Formula.}$$

$$\text{Midpoint} = \frac{2.8 + 5.6}{2} \qquad \text{Substitute values.}$$

$$\text{Midpoint} = \frac{8.4}{2} \qquad \text{Evaluate.}$$

$$\text{Midpoint} = 4.2$$

The midpoint of the segment between the points 2.8 and 5.6 is 4.2.

How to

Find the coordinates of the midpoint of a line segment if given the coordinates of the end points:

1. Average the respective coordinates of the end points of the segment using the formula

$$\text{Midpoint} = \left(\frac{x_1 + x_2}{2}, \frac{y_1 + y_2}{2} \right)$$

where P_1 and P_2 are end points of the segment, $P_1 = (x_1, y_1)$ and $P_2 = (x_2, y_2)$.
2. Substitute values for x_1, x_2, y_1, and y_2.
3. Evaluate to find the coordinates of the midpoint.

EXAMPLE   To find the midpoint of each segment with the given end points:

(a) $(2, 4)$ and $(6, 10)$ (b) $(3, -5)$ and origin

(a) $(2, 4)$ and $(6, 10)$

$$\text{Midpoint} = \left(\frac{x_1 + x_2}{2}, \frac{y_1 + y_2}{2} \right) \qquad \text{Substitute: } x_1 = 2, x_2 = 6, y_1 = 4, y_2 = 10.$$

$$\text{Midpoint} = \left(\frac{2 + 6}{2}, \frac{4 + 10}{2} \right) \qquad \text{Evaluate.}$$

$$\text{Midpoint} = \left(\frac{8}{2}, \frac{14}{2} \right)$$

$$\textbf{Midpoint} = \textbf{(4, 7)} \qquad \text{Reduce fractions.}$$

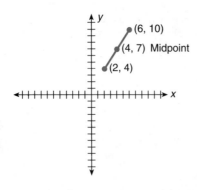

(b) $(3, -5)$ and origin.

$$\text{Midpoint} = \left(\frac{x_1 + x_2}{2}, \frac{y_1 + y_2}{2}\right)$$ Substitute: $x_1 = 3$, $x_2 = 0$, $y_1 = -5$, $y_2 = 0$.

$$\text{Midpoint} = \left(\frac{3 + 0}{2}, \frac{-5 + 0}{2}\right)$$ Evaluate.

$$\text{Midpoint} = \left(\frac{3}{2}, \frac{-5}{2}\right) \quad \text{or} \quad \left(1\frac{1}{2}, -2\frac{1}{2}\right)$$

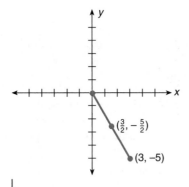

Midpoint between the origin and another point

To find the coordinates of the midpoint of a segment from the origin to any point, take one-half of each coordinate of the point that is not at the origin.

CHAPTER 10

Linear and Absolute Value Inequalities

- ■ **10–1 Inequalities and Sets**
- ■ **10–2 Solving Linear Inequalities in One Variable**
- ■ **10–3 Graphing Linear Equalities in Two Variables**
- ■ **10–4 Absolute-Value Inequalities**

■ 10–1 Inequalities and Sets

An **inequality** is a mathematical statement showing quantities that *are not equal.*

$5 \neq 7$ *5 is not equal to 7.*

To **solve an inequality,** find the value or set of values of the unknown quantity that makes the statement true.

● **RELATED TOPIC: Inequality symbols**

A **set** is a group or collection of items. For example, a set of days of the week that begin with the letter T includes Tuesday and Thursday. However, in this chapter we examine sets of numbers. The items or numbers that belong to a set are called **members** or **elements** of a set. The description of a set clearly distinguishes between the elements that belong to the set and those that do not belong. This description can be given in words or by using various types of **set notation.**

One notation is to make a **list** or **roster** of the elements of a set. These elements are enclosed in braces and separated with commas.

Set of whole numbers between 1 and 8 = {2, 3, 4, 5, 6, 7}

Another method of illustrating a set is **set-builder notation.** The elements of the set are written in the form of an inequality using a variable to represent all the elements of the set.

We can "build" the set of whole numbers between 1 and 8 with the symbolic statement

Set of whole numbers between 1 and 8 = $\{x \mid x \in W \text{ and } 1 < x < 8\}$

This statement is read "the set of values of x *such that* each x *is an element of* the set of whole numbers and x is between 1 and 8."

A special set is the empty set. The **empty set** is a set containing no elements. Symbolically, the empty set is identified as { } or φ. The symbol φ is the Greek letter phi, (pronounced "fee"). An example of an empty set is the set of whole numbers between 1 and 2.

Common symbols for sets

The following capital letters are used to denote the indicated set of numbers:

N = natural numbers	W = whole numbers
Z = integers	Q = rational numbers
I = irrational numbers	R = real numbers
M = imaginary numbers	C = complex numbers

Symbols that substitute for phrases and words that are often used in describing sets are

| is read "such that"   ∈ is read "is an element of"

How to

Graph an inequality in one variable:

1. Determine the boundaries of the inequality, if they exist.
2. Locate the boundaries on a number line.
   a. If the inequality is < or >, the boundary is *not* included. Represent this with an open circle or a parenthesis.
   b. If the inequality is ≤ or ≥, the boundary *is* included. Represent this with a solid or shaded circle or a bracket.
3. Shade the portion of the number line between the boundaries. If there is no boundary in one or either direction, the graph continues indefinitely in that direction.

Another type of notation used to represent inequalities of sets of real numbers is **interval notation.** The two **boundaries** are separated by a comma and enclosed with a symbol that indicates whether the boundary is included or not.

How to

Write an inequality in interval notation:

1. Determine the boundaries of the interval, if they exist.

2. Write the left boundary or $-\infty$ first and write the right boundary or $+\infty$ second. Separate the two with a comma.

3. Enclose the boundaries with the appropriate grouping symbols. A parenthesis (or) indicates that the boundary is *not* included. A bracket [or] indicates that a boundary *is* included.

EXAMPLE To represent the sets of numbers on the number line and by using interval notation:

  (a) $1 < x < 8$ (b) $1 \le x \le 8$ (c) $x < 3$

  (d) $x \ge 3$ (e) all real numbers

  (a) ◄─┼─⊕─┼─┼─┼─┼─┼─┼─⊕─┼─► $(1, 8)$
       0 1 2 3 4 5 6 7 8 9

   Parentheses and open circles are used for both boundaries to indicate that 1 and 8 are *not* included in the solution set.

  (b) ◄─┼─⊕─┼─┼─┼─┼─┼─┼─⊕─┼─► $[1, 8]$
       0 1 2 3 4 5 6 7 8 9

   Brackets and shaded circles are used for both boundaries to indicate that 1 and 8 *are* included in the solution set.

  (c) ◄─┼─┼─⊕─┼─┼─► $(-\infty, 3)$
       1 2 3 4 5

   The left boundary is $-\infty$, and the right boundary is 3 but is not in the solution set so a parenthesis or open circle is used.

  (d) ◄─┼─┼─⊕─┼─┼─► $[3, \infty]$
       1 2 3 4 5

   The left boundary is 3 and since 3 *is* included in the solution set, a bracket or shaded circle is used. The right boundary is ∞.

  (e) ◄─┼─┼─┼─┼─┼─┼─┼─► $(-\infty, \infty)$
       -3 -2 -1 0 1 2 3

   All real numbers is an unbounded set of numbers, thus parentheses are used. The number line extends indefinitely in both directions.

TIP

Alternate representation of inequality on a number line

Boundaries on a number line can also be indicated with parentheses or brackets.

   $-1 \le x < 2$ ◄─┼─[─┼─┼─)─►
                       -2 -1 0 1 2

Inequalities

■ 10–2 Solving Linear Inequalities in One Variable

The **sense of an inequality** is the appropriate comparison symbol: less than, greater than, less than or equal to, and greater than or equal to.

Interchanging the sides of an inequality:

When the sides of an inequality are interchanged, the sense of the inequality is reversed.

If $a < b$, then $b > a$.   If $a > b$, then $b < a$.

If $a \leq b$, then $b \geq a$.   If $a \geq b$, then $b \leq a$.

Multiplying or dividing an inequality by a negative number:

If both sides of an inequality are multiplied or divided by a negative number, the sense of the inequality is reversed.

If $a < b$, then $-a > -b$.   If $a > b$, then $-a < -b$.

If $a \leq b$, then $-a \geq -b$.   If $a \geq b$, then $-a \leq -b$.

How to

Solve linear inequalities in one variable:

1. Follow the same sequence of steps that is used to solve a similar equation.
2. The sense of the inequality remains the same unless one of the following situations occurs:
   a. The sides of the inequality are interchanged.
   b. The steps used in solving the inequality require that the entire inequality (both sides) be multiplied or divided by a negative number.
3. If either situation a or b in Step 2 occurs in solving an inequality, *reverse* the sense of the inequality; that is, less than ($<$) becomes greater than ($>$), and vice versa.

EXAMPLE To solve the inequality, $4x - 2 \le 3(25 - x)$:

$4x - 2 \le 3(25 - x)$ | Remove the parentheses.

$4x - 2 \le 75 - 3x$ | Collect letter terms on the left and number terms on the right (sort).

$4x + 3x \le 75 + 2$ | Combine like terms.

$7x \le 77$ | Divide by the coefficient of the letter.

$\dfrac{7x}{7} \le \dfrac{77}{7}$ | Symbolic, interval notation and graphical representation of the solution set.

$x \le 11$ or $(-\infty, 11]$

Inequalities

TIP

Three ways to represent the solution set of inequalities: Symbolically, graphically, and interval notation

In the preceding example, the solution to the inequality $4x - 2 \le 3(25 - x)$ was written three ways.

- Symbolic representation: $x \le 11$
- Graphical representation: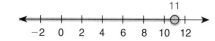
- Interval notation: $(\infty, 11]$

The symbolic representation evolves from the process of solving the inequality using the properties of inequalities and techniques for isolating the variable. Both the graphical representation and the interval notation help us interpret and visualize the solution.

● **RELATED TOPIC: Linear equations**

A **compound inequality** is a mathematical statement that combines two statements of inequality. The conditions placed on a compound inequality may use the **connective** *and* to indicate both conditions must be met simultaneously. Such compound inequalities may be written as a continuous statement. The conditions placed on a compound inequality may use the **connective** *or* to indicate that either condition may be met. Such compound inequalities *must* be written as two separate statements using the connective *or.*

A set is a **subset** of a second set if every element of the first set is also an element of the second set. If $A = \{1, 2, 3\}$ and $B = \{2\}$, then B is a subset of A. This is written in symbols as $B \subset A$

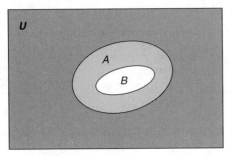

$B \subset A$

The **universal set** includes all the elements for a given description. The universal set is sometimes identified as U.

The **empty set** is a set containing no elements. The empty set is identified as $\{\ \ \}$ or ϕ.

Special set relationships:

Every set is a subset of itself: $A \subset A$

The empty set is a subset of every set: $\phi \subset A$

The **union** of two sets is a set that includes all elements that appear in *either* of the two sets. Union is generally associated with the condition "or." The symbol for union is $\cup$. The colored portion in the figure represents $A \cup B$.

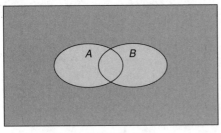

$A \cup B$

How to

Find the union of two sets:

1. Identify all the elements of each set.
2. List all elements from each set in the union. Duplicate elements from the two sets are listed in the union only once.

EXAMPLE   To find the union of the sets $A = \{2, 3, 4\}$ and $B = \{4, 5, 6\}$:

2, 3, 4, 4, 5, 6   Identify all elements of each set.

$A \cup B = \{2, 3, 4, 5, 6\}$ List all elements from each set, but list duplicate elements (4) only once.

└

The **intersection** of two sets is a set that includes all elements that appear in *both* of the two sets. Intersection is generally associated with the condition "and." The symbol for intersection is ∩. The colored portion of the figure represents $A \cap B$.

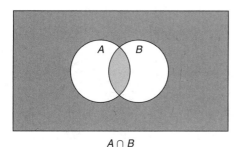

$A \cap B$

How to

Find the intersection of two sets:

1. Identify the elements from each set that appear in *both* sets.
2. List the elements from Step 1 in the intersection set.

EXAMPLE   To find the intersection of the two sets, $A = \{-1, 2, 3, 4\}$ and $B = \{2, 3, 4, 5, 6\}$:

$-1, \underline{2, 3, 4}$ Identify the elements from each set that appear in
  $\underline{2, 3, 4,} 5, 6$ both sets.

$A \cap B = \{2, 3, 4\}$ List the elements that appear in both sets as the
                intersection of sets.

⌞

A **conjunction** is an intersection or **"and"** set relationship.

Property for conjunctions:

If $a < x$ and $x < b$, then $a < x < b$.

If $a > x$ and $x > b$, then $a > x > b$.

Similar compound inequalities may also use $\leq$ and $\geq$.

If $a \leq x$ and $x \leq b$, then $a \leq x \leq b$.

If $a \geq x$ and $x \geq b$, then $a \geq x \geq b$.

How to

Solve a compound inequality that is a conjunction:

1. Separate the compound inequality into two simple inequalities using the conditions of the conjunction.
2. Solve each simple inequality.
3. Determine the solution set that includes the *intersection* of the solution sets of the two simple inequalities.

EXAMPLE   To find the solution set for each compound inequality, graph the solution set on a number line, and write the solution set in interval notation:

(a) $5 < x + 3 < 12$ (b) $-7 \leq x - 7 \leq 1$

(a) $5 < \boxed{x + 3} < 12$ Separate into two simple inequalities.

$$5 < \boxed{x + 3} \quad \text{and} \quad \boxed{x + 3} < 12 \qquad \text{Solve each inequality.}$$

$$5 - 3 < x \qquad\qquad x < 12 - 3$$

$$2 < x \quad \text{or} \quad x > 2 \qquad x < 9 \qquad \text{Find overlap.}$$

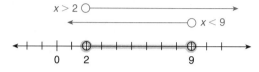

The solution set is a continuous statement $2 < x < 9$ and in interval notation, $(2, 9)$.

(b) $-7 \leq \boxed{x - 7} \leq 1$ Separate into two simple inequalities.

$$-7 \leq \boxed{x - 7} \quad \text{and} \quad \boxed{x - 7} \leq 1 \qquad \text{Solve each inequality.}$$

$$-7 + 7 \leq x \qquad\qquad x \leq 1 + 7$$

$$0 \leq x \quad \text{or} \quad x \geq 0 \qquad x \leq 8 \qquad \text{Find overlap.}$$

The figure shows the solution set graphically. The solution set as a continuous statement is $0 \leq x \leq 8$ and in interval notation, $[0, 8]$.

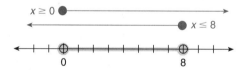

TIP

Greater than versus less than

Even though continuous compound inequalities can be written using greater than or less than, using less than follows the natural positions of the boundaries on the number line. If $a > b > c$, then $c < b < a$. For instance, if $3 > 0 > -2$, then $-2 < 0 < 3$. Compound inequalities using the "less than" symbols are used most often.

A **disjunction** is a union or "**or**" set relationship. Either condition is met in a disjunction. A disjunction *cannot* be written as a continuous statement.

How to

Solve a compound inequality that is a disjunction:

1. Solve each simple inequality.
2. Determine the solution set that includes the union of the solution sets of the two simple inequalities.

EXAMPLE   To find the solution set for each compound inequality, graph the solution set on a number line, and write the solution in interval notation:

(a) $x + 3 < -2$ or $x + 3 > 2$

(b) $x - 5 \le -3$ or $x - 5 \ge 3$

(a) $x + 3 < -2$ or $x + 3 > 2$ Solve each simple inequality.

   $x < -2 - 3$ $x > 2 - 3$

   $x < -5$ or $x > -1$ Solution set.

   $(-\infty, -5)$ or $(-1, \infty)$

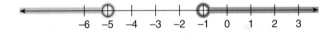

(b) $x - 5 \le -3$ or $x - 5 \ge 3$ Solve each simple inequality.

   $x \le -3 + 5$ $x \ge 3 + 5$

   $x \le 2$ or $x \ge 8$ Solution set.

   $(-\infty, 2]$ or $[8, \infty)$

■ 10–3 Graphing Linear Inequalities in Two Variables

The solution of a linear inequality in two variables is the set of ordered pairs that make a true statement when substituted in the linear equality. This solution set is represented by a shaded portion of a rectangular coordinate system. The boundary of the shaded portion is the graph of the linear equation that replaces the inequality symbol with an equal symbol.

How to

Graph a linear inequality with two variables:

1. Find the boundary by graphing an equation that substitutes an equal sign for the inequality symbol. Make a solid line if the boundary is included ($\leq$ or $\geq$) or a dashed line if the boundary is not included ($<$ or $>$) in the solution set.
2. Test any point that is not on the boundary line.
3. If the test point makes a true statement with the original inequality, shade the side containing the test point.
4. If the test point makes a false statement with the original inequality, shade the side opposite the side containing the test point.

EXAMPLE   Graph the inequality $2x + y < 3$.
Graph the equation $2x + y = 3$ to establish the boundary. The boundary will *not* be included in the solution set.

$$2x + y = 3 \qquad \text{Solve the equation for } y.$$

$$y = -2x + 3 \qquad \text{Identify the slope and } y\text{-intercept.}$$

$$\text{slope} = -2 \text{ or } \frac{-2}{1} \text{ or } \frac{2}{-1} \qquad \text{Coefficient of } x.$$

$$y\text{-intercept} = 3 \text{ or } (0, 3) \qquad \text{Constant.}$$

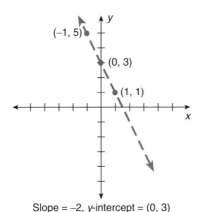

Slope = –2, y-intercept = (0, 3)

To show that the boundary is not included in the solution set, a dashed or broken line is used. The graph of the equation $y = -2x + 3$ divides the surface into two parts and it is the *boundary* separating the parts. In graphing inequalities, one of the parts is included in the solution. To represent this solution, we *shade* the appropriate part.

Inequalities

Now, select one point, on either side of the boundary, to determine which side of the line contains the set of points that solves the inequality. Suppose that we choose (2, 2) as the point above the boundary.

Is $2x + y < 3$ when $x = 2$ and $y = 2$?

$2(2) + 2 < 3$

$4 + 2 < 3$

$6 < 3$ False statement.

Thus, the side of the line including the point (2, 2) is *not* in the solution set. Shade the opposite side.

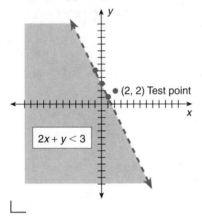

TIP

Solid or dashed boundary lines

In inequalities with one variable, we use solid circles and open circles to show that boundaries are included or excluded. In inequalities with two variables, we represent boundaries that are included in the solution with solid lines and boundaries that are excluded from the solution with dashed lines. The graph shows $2x + y \leq 3$.

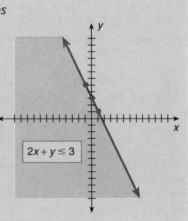

TIP

Selecting test points

Choose numbers that are easy to work with when selecting a test point. If the boundary line does not pass through the origin, a good point to use is $(0, 0)$.

■ 10–4 Absolute-Value Inequalities

"Less than" relationships for absolute-value inequalities:

    If $|x| < b$ and $b > 0$, then $-b < x < b$ or $(-b, b)$.

                     and

    If $|x| \leq b$ and $b > 0$, then $-b \leq x \leq b$ or $[-b, b]$.

Absolute-value inequalities with the "less than" or "less than or equal to" relationship have a solution set that is a **conjunction (and)**. The graph of the solution is one continuous interval.

$-b \leq x \geq b$ or $[-b, b]$

How to

Solve an absolute-value inequality with the "less than" inequality:

1. Isolate the absolute-value term.
2. Write the absolute-value inequality as a compound inequality that is a conjunction.
3. Solve the compound inequality.

EXAMPLE   To find the solution set for each inequality:

    (a) $|x| < 3$ (b) $|x + 5| < 8$ (c) $|x - 1| \leq -5$

    (d) $|3x| + 2 \leq 14$

Inequalities

(a) $|x| < 3$ Apply the property of absolute-value inequalities having $<$ relationship.

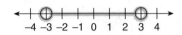

$-3 < x < 3$ or $(-3, 3)$.

(b) $|x + 5| < 8$ Apply the property of absolute-value inequalities having $<$ relationship.

$-8 < \boxed{x + 5} < 8$ Separate into two simple inequalities.

$-8 < \boxed{x + 5}$ or $\boxed{x + 5} < 8$ Solve each inequality.

$-8 - 5 < x$ $x < 8 - 5$

$-13 < x$ or $x > -13$ $x < 3$ Find overlap.

$-13 < x < 3$ or $(-13, 3)$.

(c) $|x - 1| \le -5$ This is a special case of absolute-value *less than* inequalities.

The absolute value of an expression is nonnegative. So $|x - 1| \le -5$ has no solution.

(d) $|3x| + 2 \le 14$ Isolate the absolute-value term.

$|3x| \le 14 - 2$

$|3x| \le 12$ Apply the property of absolute-value inequalities having $\le$ relationship.

$-12 \le \boxed{3x} \le 12$ Separate into two simple inequalities.

$-12 \le \boxed{3x}$ or $\boxed{3x} \le 12$ Divide.

$-4 \le x$ $x \le 4$ Find overlap.

$-4 \le x \le 4$ or $[-4, 4]$ is the solution set.

"Greater than" relationships for absolute-value inequalities:

If $|x| > b$ and $b > 0$, then $x < -b$ or $x > b$.

or

If $|x| \geq b$ and $b > 0$, then $x \leq -b$ or $x \geq b$.

Absolute-value inequalities with the "greater than" or "greater than or equal to" relationship have a solution set that is a **disjunction (or).** The graph is two separate continuous intervals.

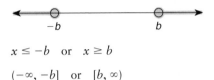

$x \leq -b$ or $x \geq b$

$(-\infty, -b]$ or $[b, \infty)$

How to

Solve an absolute-value inequality with the "greater-than" inequality:

1. Isolate the absolute-value term.
2. Write the absolute-value inequality as a compound inequality that is a disjunction.
3. Solve the compound inequality.

EXAMPLE To find the solution set for each inequality.

(a) $|x| > 5$ (b) $|x| + 1 \geq 4$ (c) $|2x - 2| \geq 7$

(d) $|x - 1| \geq -5$

(a) $|x| > 5$ Apply the property of absolute-value inequalities having > relationship.

$x < -5$ or $x > 5$

$x < -5$ or $x > 5; (-\infty, -5)$ or $(5, \infty)$.

(b) $|x| + 1 \geq 4$ Isolate the absolute-value term.

$|x| \geq 4 - 1$

$|x| \geq 3$ Apply the property of inequalities having $\geq$ relationship.

$|x| \leq -3$ or $|x| \geq 3$

$x \leq -3$ or $x \geq 3$

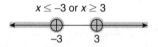

$x \leq -3$ or $x \geq 3;\ (-\infty, -3]$ or $[3, \infty)$.

(c) $|2x - 2| \geq 7$ Apply the property of absolute-value inequalities having $\geq$ relationship.

$2x - 2 \leq -7$ or $2x - 2 \geq 7$

$2x \leq -7 + 2$ $\qquad 2x \geq 7 + 2$

$2x \leq -5$ $\qquad\qquad 2x \geq 9$

$x \leq -\dfrac{5}{2}$ $\qquad\qquad x \geq \dfrac{9}{2}$

$x \leq -\frac{5}{2}$ or $x \geq \frac{9}{2}$

Thus, $x \leq -\frac{5}{2}$ or $x \geq \frac{9}{2}, (-\infty, -\frac{5}{2}]$ or $[\frac{9}{2}, \infty)$.

(d) $|x - 1| \geq -5$ This is a special case of absolute value *less than* inequalities.

The absolute value of an expression is nonnegative. So $|x - 1| \geq -5$ can be changed to $|x - 1| \geq 0$.

Then, apply the property of inequalities having $\geq$ relationship.

$x - 1 \leq 0$ or $x - 1 \geq 0$

$x \leq 1 + 0$ $\qquad\quad x \geq 1 + 0$

$x \leq 1$ or $x \geq 1$

$(-\infty, 1]$ or $[1, \infty)$

The absolute value of *any* expression is greater than *any* negative number. **The solution is all real numbers.** $(-\infty, \infty)$

● **RELATED TOPICS: Absolute-value linear equation, Compound inequalities, Conjunctions, Disjunctions**

CHAPTER 11

Systems of Linear Equations and Inequalities

- ■ **11–1 Solving Systems of Linear Equations**
- ■ **11–2 Solving Systems of Linear Inequalities Graphically**

■ 11–1 Solving Systems of Linear Equations

A **system of two linear equations,** having two variables each, is solved when we find ordered pairs of solutions that satisfies *both* equations.

How to

Solve a system of two equations with two variables by graphing:

1. Graph each equation on the same pair of axes.
2. The solution of the system is the coordinates of the common point or points.

EXAMPLE   To solve the system:

$$x + y = 5$$

$$x - y = 3$$

Make a table of values for each equation by solving for y.

$$x + y = 5 \qquad x - y = 3$$

$$y = 5 - x \qquad -y = 3 - x$$

$$y = x - 3$$

x	$5 - x$	y	
2	$5 - 2$	3	(2, 3)
5	$5 - 5$	0	(5, 0)

x	$x - 3$	y	
3	$3 - 3$	0	(3, 0)
5	$5 - 3$	2	(5, 2)

Graph each equation on the same grid.

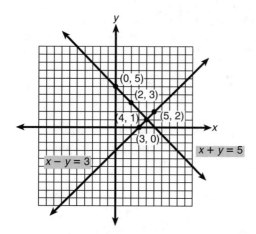

The lines cross at (4, 1), so the solution is $x = 4, y = 1$.

⌞

● **RELATED TOPICS: Graphing linear equations, Table of values**

How to

Solve a system of equations using the addition method (elimination):

1. Write each equation in standard form.
2. If necessary, multiply one or both equations by numbers that cause the terms of one variable to add to zero.
3. Add the two equations to eliminate a variable.
4. Solve the equation from Step 3.
5. Substitute the solution from Step 4 in either equation and solve for the remaining variable.
6. Check the solution in each original equation.

EXAMPLE   To solve the system of equations $x + y = 5$ and $x - y = 3$ using the elimination method:

$x + y = 5$	Equations already in standard form.
$\underline{x - y = 3}$	The y variables will eliminate. Add the equations.
$2x \quad = 8$	Solve for x.
$\boxed{x = 4}$	x-value of solution.
$\boxed{x} + y = 5$	Substitute in $x = 4$ into either original equation.
$\boxed{4} + y = 5$	Solve for y.
$y = 5 - 4$	
$\boxed{y = 1}$	y-value of solution.
$\boxed{x} + \boxed{y} = 5$	Check $x = 4$ and $y = 1$ in first original equation.
$\boxed{4} + \boxed{1} = 5$	
$5 = 5$	Solution checks.
$\boxed{x} - \boxed{y} = 3$	Check in second original equation.
$\boxed{4} - \boxed{1} = 3$	
$3 = 3$	Solution checks.

The solution is $x = 4$ and $y = 1$ or $(4, 1)$.

EXAMPLE   To solve the system of equations $2x + y = 7$ and $x + y = 3$:

$-2x - y = -7$	Multiply the first equation by -1 to make the coefficients of y opposites.
$\underline{x + y = \quad 3}$	Add the equations to eliminate the y variable.
$-x \quad = -4$	Solve the basic equation.
$x = \boxed{4}$	
$-2\boxed{x} - y = -7$	Substitute 4 in place of x in the first equation and solve for y.
$-2(\boxed{4}) - y = -7$	Multiply $-2(4)$.
$-8 - y = -7$	Add 8 to both sides and combine like terms.
$-y = -7 + 8$	Divide both sides by -1.

Systems

$$-y = 1$$
$$y = -1$$

The solution is $(4, -1)$.

Check the solution in the first original equation.

$$2x + y = 7$$
$$2(4) + (-1) = 7$$
$$8 + (-1) = 7$$
$$7 = 7$$

Check the solution in the second original equation.

$$x + y = 3$$
$$4 + (-1) = 3$$
$$3 = 3$$

Dependent equations—All solutions in common

When solving a system of equations, if both variables are eliminated and the resulting statement is true, then the equations are **dependent** and have many solutions. The graphs of the equations coincide.

Inconsistent equations—No solution

When solving a system of equations, if both variables are eliminated and the resulting statement is false, the equations are *inconsistent* and have no solution. The graphs of the equations are parallel lines.

How to

Solve a system of equations by substitution:

1. Rearrange either equation to isolate one variable.
2. Substitute the equivalent expression from Step 1 into the *other* equation and solve for the remaining variable.

3. Substitute the solution from Step 2 into the equation from Step 1 to find the value of the substituted variable.

4. Check both original equations.

EXAMPLE   To solve the system of equations using the substitution method:

$$2x - 3y = -14$$

$$x + 5y = 19$$

When using the substitution method, either equation can be solved for either unknown. In this example, the x-term in the second equation has a coefficient of 1, so the simplest choice would be to solve the second equation for x.

Step 1	Step 2	Step 3
$x + 5y = 19$	$2x - 3y = -14$	$x = 19 - 5y$
$x = 19 - 5y$	$2(19 - 5y) - 3y = -14$	$x = 19 - 5(4)$
	$38 - 10y - 3y = -14$	$x = 19 - 20$
	$38 - 13y = -14$	$x = -1$
	$-13y = -14 - 38$	
	$-13y = -52$	
	$\dfrac{-13y}{-13} = \dfrac{-52}{-13}$	
	$y = 4$	

The solution is $(-1, 4)$.

Check the roots $x = -1$, $y = 4$ in both original equations.

Step 4

$$2x - 3y = -14 \qquad\qquad x + 5y = 19$$

$$2(-1) - 3(4) = -14 \qquad -1 + 5(4) = 19$$

$$-2 - 12 = -14 \qquad\qquad -1 + 20 = 19$$

$$-14 = -14 \qquad\qquad\quad 19 = 19$$

● **RELATED TOPICS: Graphing linear equations, Parallel lines**

Systems

▲ APPLICATION: Problem Solving Using a System of Equations

> **How to**
>
> *Use a system of equations to solve a problem:*
>
> 1. Identify the two unknown variables. Identify the known quantities.
> 2. Write two equations that relate the two unknowns.
> 3. Solve the system of equations for both variables.

EXAMPLE   Rosita has $5,500 to invest and for tax purposes wants to earn exactly $500 interest for 1 year. She wants to invest part at 10% and the remainder at 5%. To find the amount she must invest at each interest rate to earn exactly $500 interest in 1 year:

Let x = the amount invested at 10%. Let y = the amount invested at 5%. Interest for one year = rate × amount invested. Remember to convert percents to decimals. Using these relationships, we derive a system of equations.

Known facts	Total of $5,500 to be invested $500 interest to be earned in one year
Unknown facts	How much should be invested at 10%? How much should be invested at 5%?
Relationships	Amount invested at 10%: x Interest earned at 10%: $0.1x$ Amount invested at 5%: y Interest earned at 5%: $0.05y$

$$x + \quad y = 5{,}500 \qquad \text{Equation 1 (total investment).}$$

$$0.1x + 0.05y = 500 \qquad \text{Equation 2 (total interest in one year).}$$

Estimation	If the total amount was invested at 10%, the interest (in one year) would be $550 (0.1 × $5,500). Since we want $500 in interest, most of the money will need to be invested at 10%.
Calculations	Solve by the substitution method.

$$x + y = 5{,}500 \qquad \text{Solve Equation 1 for } x.$$

$$x \quad = 5{,}500 - y$$

Substitute into Equation 2.

$$0.1\,x + 0.05y = 500$$

Substitute $5{,}500 - y$ for x and distribute.

$$0.1(\,5{,}500 - y\,) + 0.05y = 500$$

$$550 - 0.1y + 0.05y = 500$$

Combine like terms. $-0.10y + 0.05y = -0.05y$

$$550 - 0.05y = 500$$

Sort terms.

$$-0.05y = 500 - 550$$

$$-0.05y = -50$$

Divide.

$$y = \frac{-50}{-0.05}$$

$$y = \$1{,}000 \text{ at } 5\%$$

Substitute $\$1{,}000$ for y in Equation 1.

$$x + y = \$5{,}500$$

$$x + 1{,}000 = 5{,}500$$

$$x = 5{,}500 - 1{,}000$$

$$x = \$4{,}500 \text{ at } 10\%$$

Interpretation   **Rosita must invest \$4,500 at 10% and \$1,000 at 5% for 1 year to earn \$500 interest.**

⌞

■ 11–2 Solving Systems of Linear Inequalities Graphically

How to

Solve a system of two linear inequalities with two variables by graphing:

1. Graph each inequality on the same pair of axes.
2. The solution set of the system will be the overlapping portion of the solution sets of the two inequalities.

Systems

EXAMPLE   To solve the system of inequalities $y \leq 3x + 5$ and $x + y > 7$: Graph the inequality $y \leq 3x + 5$. The boundary $y = 3x + 5$ has a slope of 3 and a y-intercept of 5. The boundary will be included in the solution set.

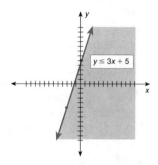

Because $(0, 0)$ does not fall on the graph of $y = 3x + 5$, it is used as the test point in the original inequality, $y \leq 3x + 5$.

For $(0, 0)$

$y \leq 3x + 5$

$0 \leq 3(0) + 5$

$0 \leq 0 + 5$

$0 < 5$ True.

The point $(0, 0)$ is included in the solution set.

Graph the inequality, $x + y > 7$. The boundary $x + y = 7$ in slope-intercept form is $y = -x + 7$. The slope is -1 and the y-intercept is 7. ($x + y = 7$ is not included in the solution set.)

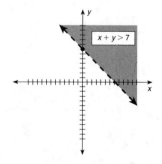

The point $(0, 0)$ does not fall on the graph of $x + y = 7$. Use it as the test point in the original inequality $x + y > 7$.

For (0, 0)

$x + y > 7$

$0 + 0 > 7$

$0 > 7$ False.

The point (0, 0) is not included in the solution set. The solution set will be the opposite side of the boundary.

Find the solution set common to both inequalities. Visualize both graphs on the same axes. The portion that has overlapping shading represents the points that satisfy *both* conditions. The shaded area forms the solution set for the system of inequalities.

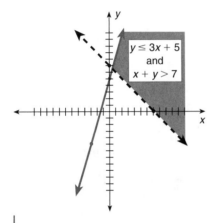

● **RELATED TOPICS: Graphing linear equalities, Graphing linear inequalities**

Powers and Polynomials

■ 12–1 Laws of Exponents

> **How to**
>
> *Multiply powers that have like bases:*
>
> 1. Verify that the bases are the same. Use this base as the base of the product.
> 2. Add the exponents for the exponent of the product.
>
> This rule can be stated symbolically as
>
> $$a^m(a^n) = a^{m+n} \qquad \text{where } a, m, \text{ and } n \text{ are real numbers.}$$

EXAMPLE   To write the products:

(a) $y^4(y^3)$ (b) $a(a^2)$ (c) $b(b)$

(a) $y^4(y^3) = y^{4+3} = y^7$ The bases are the same, so add the exponents.

(b) $a(a^2) = a^{1+2} = a^3$ When the exponent for a base is not written, it is 1.

(c) $b(b) = b^{1+1} = b^2$

Dividing by a number is equivalent to multiplying by the reciprocal of the number. Dividing powers with like bases introduces a new notation for reciprocals.

TIP

Reciprocals and negative exponents

A number times its reciprocal equals one.

$$n \cdot \frac{1}{n} = 1 \qquad n^1 \cdot n^{-1} = n^0 = 1$$

An expression with a **negative exponent** can be written as an equivalent expression with a positive exponent.

$$n^{-1} = \frac{1}{n} \qquad \frac{1}{n^{-1}} = n$$

How to

Divide powers that have like bases:

1. Verify that the bases are the same. Use this base as the base of the quotient.
2. Subtract the exponents for the exponent of the quotient.

This rule can be stated symbolically as

$$\frac{a^m}{a^n} = a^{m-n}$$

where a, m, and n are real numbers except that $a \neq 0$.

Powers and Polynomials

● **RELATED TOPICS:** Dividing fractions, Zero exponents

EXAMPLE   To write the quotients using positive exponents:

(a) $\dfrac{x^5}{x^8}$ (b) $\dfrac{a}{a^4}$ (c) $\dfrac{y^{-3}}{y^2}$ (d) $\dfrac{x^3}{x^{-5}}$

(a) $\dfrac{x^5}{x^8} = x^{5-8} = x^{-3} = \dfrac{1}{x^3}$ (b) $\dfrac{a}{a^4} = a^{1-4} = a^{-3} = \dfrac{1}{a^3}$

(c) $\dfrac{y^{-3}}{y^2} = y^{-3-2} = y^{-5} = \dfrac{1}{y^5}$ (d) $\dfrac{x^3}{x^{-5}} = x^{3-(-5)} = x^{3+5} = x^8$

How to

Raise a power to a power:

1. Multiply exponents.
2. Keep the same base.

This rule can be stated symbolically as

$$(a^m)^n = a^{mn}$$

where a, m, and n are real numbers.

EXAMPLE   To raise the power to the indicated power for:

(a) $(3^2)^3$ (b) $(x^3)^4$

(a) $(3^2)^3 = 3^{2(3)} = 3^6 = 729$

(b) $(x^3)^4 = x^{3(4)} = x^{12}$

How to

Raise a fraction or quotient to a power:

1. Raise the numerator to the power.
2. Raise the denominator to the power.

$$\left(\frac{a}{b}\right)^n = \frac{a^n}{b^n} \qquad b \neq 0$$

EXAMPLE   To raise the fractions to the indicated power:

(a) $\left(\frac{2}{3}\right)^2$ (b) $\left(\frac{-3}{4}\right)^2$ (c) $\left(\frac{x^2}{y^3}\right)^4$

(a) $\left(\frac{2}{3}\right)^2 = \frac{2^2}{3^2} = \frac{4}{9}$

(b) $\left(\frac{-3}{4}\right)^2 = \frac{(-3)^2}{4^2} = \frac{9}{16}$ $(-3)(-3) = +9$

(c) $\left(\frac{x^2}{y^3}\right)^4 = \frac{(x^2)^4}{(y^3)^4} = \frac{x^8}{y^{12}}$

How to

Raise a product to a power:

Raise each factor to the indicated power.

$(ab)^n = a^n b^n$

EXAMPLE   Raise the products to the indicated powers.

   (a) $(ab)^2$ (b) $(a^2b)^3$ (c) $(3x)^2$ (d) $(-5xy)^2$

   (a) $(ab)^2 = (a^1b^1)^2 = a^{1(2)}b^{1(2)} = a^2b^2$

   (b) $(a^2b)^3 = a^{2(3)}b^{1(3)} = a^6b^3$

   (c) $(3x)^2 = 3^{1(2)}x^{1(2)} = 3^2x^2 = 9x^2$

   (d) $(-5xy)^2 = (-5)^{1(2)}x^{1(2)}y^{1(2)} = (-5)^2x^2y^2 = 25x^2y^2$

◼ 12–2 Basic Operations with Algebraic Expressions

Like terms are terms that have the same variables and exponents.

How to

Combine like terms:

1. Combine the coefficients using the rules for adding or subtracting signed numbers.
2. The letter factors and exponents do not change.

EXAMPLE   To simplify the algebraic expressions by combining like terms:

   (a) $5x^3 + 2x^3$ (b) $x^5 - 4x^5$ (c) $a^3 + 4a^2 + 3a^3 - 6a^2$

   (a) $5\,x^3 + 2\,x^3 = (5 + 2)\,x^3 = 7\,x^3$

   $5x^3$ and $2x^3$ are like terms; therefore, add the coefficients 5 and 2. The sum has the same letter factor and exponent as the like terms.

(b) $\boxed{x^5} - 4\boxed{x^5} = \boxed{(1-4)}\boxed{x^5} = \boxed{-3}\boxed{x^5}$

x^5 and $-4x^5$ are like terms. $1 - 4 = -3$. The sum has the same letter factor and exponent as the like terms.

(c) $\boxed{a^3} + \boxed{4a^2} + 3a^3 \boxed{- 6a^2} = \boxed{4a^3} \boxed{- 2a^2}$

a^3 and $3a^3$ are like terms. $4a^2$ and $-6a^2$ are like terms. Combine coefficients mentally.

Is combining like terms the same as adding? And, what does simplify mean?

■ Combining versus adding

With the introduction of integers into our number system, we broaden our concept of addition to include both addition and subtraction. That is, when adding integers with like signs, we add absolute values, and when adding integers with unlike signs, we subtract absolute values. Also, we developed a strategy for interpreting any subtraction as an equivalent addition by changing the subtrahend to its opposite.

It is common to use the expression **combine terms** when referring to addition or to subtraction of signed numbers.

■ Simplifying

The instructions to **simplify the expression** are vague but often used in mathematics exercises. In general, to simplify an expression means to write the expression using fewer terms or reduced or with lower coefficients or exponents. When the instructions say "simplify," the intent is for you to examine the expression and see which laws or operations allow you to rewrite the expression in a simpler form.

How to

Multiply or divide algebraic expressions:

1. Multiply or divide the coefficients using the rules for signed numbers.
2. Multiply or divide the letter factors using the laws of exponents for factors with like bases.

EXAMPLE To multiply or divide and express answers with positive exponents:

(a) $(4x)(3x^2)$ (b) $(-6y^2)(2y^3)$ (c) $\dfrac{2x^4}{x}$ (d) $\dfrac{-5x^4}{15x^2}$

(a) $(4x)(3x^2) = 4(3)\ (x^{1+2}) = 12\ x^3$
   Multiply coefficients. Add exponents.

(b) $(-6y^2)(2y^3) = -6(2)\ (y^{2+3}) = -12\ y^5$
   Multiply coefficients. Add exponents.

(c) $\dfrac{2x^4}{x} = \dfrac{2}{1}\ (x^{4-1}) = 2\ x^3$

   Divide coefficients. Subtract exponents. The coefficient of x in the denominator is 1.

(d) $\dfrac{-5x^4}{15x^2} = \dfrac{-5}{15}\ (x^{4-2}) = -\dfrac{1}{3}x^2$ or $-\dfrac{x^2}{3}$

   Reduce coefficients. Subtract the exponents of like base x.
   $-\dfrac{1}{3}x^2$ is the same as $-\dfrac{1}{3}\left(\dfrac{x^2}{1}\right)$ or $-\dfrac{x^2}{3}$.

TIP

Why can't we cancel terms?

Reduce or cancel *factors*, not *terms*. Look at

$$\frac{6x^3 + 2x^2}{2x^2}$$

A common *mistake* is to cancel the terms.

$$\frac{6x^3 + 2x^2}{2x^2} = 6x^3 \qquad \text{Not correct.}$$

Why is this not correct? To check a division, multiply the quotient by the divisor (denominator). The result will be the dividend (numerator).

$6x^3(2x^2) = 6x^3 + 2x^2$ is *not* true.

● **RELATED TOPICS: Multiplying fractions, Reducing fractions**

Powers and
Polynomials

■ 12–3 Powers of 10 and Scientific Notation

• Decimal point

| $1{,}000{,}000$ or 10^6 | $100{,}000$ or 10^5 | $10{,}000$ or 10^4 | $1{,}000$ or 10^3 | 100 or 10^2 | 10 or 10^1 | 1 or 10^0 | $\frac{1}{10}$ or 10^{-1} | $\frac{1}{100}$ or 10^{-2} | $\frac{1}{1{,}000}$ or 10^{-3} | $\frac{1}{10{,}000}$ or 10^{-4} | $\frac{1}{100{,}000}$ or 10^{-5} | $\frac{1}{1{,}000{,}000}$ or 10^{-6} |

Base-ten place-value chart.

TIP

What does the exponent in a power of 10 tell us?

The absolute value of the exponent equals the *number of zeros* in the ordinary number in whole-number or fractional form.

We learned in arithmetic that we can multiply or divide by 10, 100, 1,000, and so on, by shifting the decimal point. When we are multiplying or dividing by a power of 10, **the exponent of 10 tells how many places the decimal is to be shifted and in what direction.**

Positive exponent *implies* shift right. Negative exponent *implies* shift left.

How to

Multiply by a power of 10:

1. If the exponent is positive, shift the decimal point to the *right* the number of places indicated by the *positive* exponent. Attach zeros as necessary.
2. If the exponent is negative, shift the decimal point to the *left* the number of places indicated by the *negative* exponent. Insert zeros as necessary.

EXAMPLE   To multiply by powers of 10:

(a) 275×10^3 (b) 2.4×10^{-2} (c) 43×10^{-1}

(a) 275×10^3 The exponent is positive; shift the decimal 3 places to
     275,000   the right. Attach 3 zeros.

(b) 2.4×10^{-2} The exponent is negative; shift the decimal 2 places
     0.024   to the left. Insert 1 zero.

(c) 43×10^{-1} The exponent is negative; shift the decimal 1 place to
     4.3   the left.

⌞

How to

Divide by a power of 10:

1. Change the division to an equivalent multiplication.
2. Use the rule for multiplying by a power of 10.

● **RELATED TOPICS: Dividing fractions, Negative exponents**

EXAMPLE   To divide by powers of 10:

(a) $3.14 \div 10$ (b) $0.48 \div 10^2$ (c) $20.1 \div 10^{-2}$

(a) $3.14 \div 10 = 3.14 \times \frac{1}{10}$ Change the division to an
                                                       equivalent multiplication.

$\qquad = 3.14 \times 10^{-1}$ Express the fraction $\frac{1}{10}$ as a power of
                                       10.

$\qquad = \mathbf{0.314}$ Because the exponent of 10 is -1,
                                       move the decimal one place to the
                                       left.

(b) $0.48 \div 10^2 = 0.48 \times \frac{1}{100}$

$\qquad = 0.48 \times 10^{-2}$

$\qquad = \mathbf{0.0048}$ Move decimal two places to the left.
                                       Insert two zeros.

(c) $20.1 \div 10^{-2} = 20.1 \times 10^2$ Change division to equivalent
                                                    multiplication.

$\qquad = \mathbf{2{,}010}$ Shift decimal two places to the right.

⌞

A number is expressed in **scientific notation** if it is the product of two
factors. The absolute value of the first factor is a number greater than or
equal to 1 but less than 10. The second factor is a power of 10.

A number that is written strictly according to place value is an **ordinary number.**

Characteristics of scientific notation

- Numbers between 0 and 1 and between −1 and 0 require negative exponents when written in scientific notation.
- The first factor in scientific notation always has only one nonzero digit to the left of the decimal.
- The use of the times sign (×) for multiplication is the most common representation for scientific notation.

How to

Change from a number written in scientific notation to an ordinary number:

1. Perform the indicated multiplication by moving the decimal point in the first factor the appropriate number of places.
2. Affix or insert zeros as necessary.

EXAMPLE   To change to ordinary numbers:

(a) 3.6×10^4 (b) 2.8×10^{-2} (c) 1.1×10^0

(d) 6.9×10^{-5} (e) 9.7×10^6

(a) $3.6 \times 10^4 = 36000. = 36,000$ Perform the indicated operation. Move the decimal four places to the right.

(b) $2.8 \times 10^{-2} = .028 = 0.028$ Move the decimal two places to the left.

(c) $1.1 \times 10^0 = 1.1$ Move the decimal no (zero) places.

(d) $6.9 \times 10^{-5} = .000069 = 0.000069$ Move the decimal five places to the left.

(e) $9.7 \times 10^6 = 9700000. = 9,700,000$ Move the decimal six places to the right.

How to

Change from a number written in ordinary notation to scientific notation:

1. Indicate where the decimal should be positioned in the ordinary number so that the absolute value of the number is valued at 1 or between 1 and 10 by inserting a caret ($\wedge$) in the proper place.
2. Determine how many places and in which direction the decimal shifts *from* the new position (caret) *to* the old position (original decimal point). This number is the exponent of the power of 10.

EXAMPLE   To express in scientific notation:

   (a)   285   (b)   0.007   (c)   9.1   (d)   85,000

   (a) $285 \rightarrow 2_\wedge 85 = 2.85 \times 10^2$

      The unwritten decimal is after the 5. Place the caret between 2 and 8 so the number 2.85 is between 1 and 10. Count *from* the caret *to* the decimal to determine the exponent of 10. A move two places to the right represents the exponent +2.

   (b) $0.007 \rightarrow 0.\,007_\wedge = 7 \times 10^{-3}$

      7 is between 1 and 10. Count *from* the caret *to* the decimal. A move three places to the left represents the exponent −3.

   (c) $9.1 = 9.1 \times 10^0$

      9.1 is already between 1 and 10, so the decimal does not move; that is, the decimal moves zero places.

   (d) $85,000 \rightarrow 8_\wedge 5000 = 8.5 \times 10^4$

      From the caret *to* the decimal is four places to the right.

How to

Multiply numbers in scientific notation:

1. Multiply the first factors using the rules of signed numbers.
2. Multiply the power-of-10 factors using the laws of exponents.
3. Examine the first factor of the product (Step 1) to see if its value is equal to 1 or between 1 and 10.
    a. If so, write the results of Steps 1 and 2.
    b. If not, shift the decimal so the first factor is equal to 1 or is between 1 and 10, and adjust the exponent of the power-of-10 factor accordingly.

Powers and Polynomials

TIP

A balancing act: Why count from the new to the old?

Moving the decimal in the ordinary number changes the value of the number unless you balance the effect of the move in the power-of-10 factor. When a decimal is moved in the first factor, the value is changed. To offset this change, an opposite change must be made in the power-of-10 factor. From the new to the old position indicates the proper number of places and the direction (positive or negative) for balancing with the power-of-10 factor.

Remember the word *NO*. Count from N ew to O ld.

$$3,800 = 3.8 \times 10^3 \qquad 3_\wedge 800. \times 10^3 \qquad N \to O = +3$$
$$0.0045 = 4.5 \times 10^{-3} \qquad 0.004_\wedge 5 \times 10^{-3} \qquad N \to 0 = -3$$

EXAMPLE To multiply:

(a) $(4 \times 10^2)(2 \times 10^3)$ (b) $(3.7 \times 10^3)(2.5 \times 10^{-1})$

(c) $(8.4 \times 10^{-2})(5.2 \times 10^{-3})$

(a) $(4 \times 10^2)(2 \times 10^3) = 8 \times 10^5$

Because 8 is between 1 and 10. Adjustments are not necessary.

(b) $(3.7 \times 10^3)(2.5 \times 10^{-1}) = 9.25 \times 10^2$

Because 9.25 is between 1 and 10. Adjustments are not necessary.

(c) $(8.4 \times 10^{-2})(5.2 \times 10^{-3}) = 43.68 \times 10^{-5}$

43.68 is not between 1 and 10. Adjustments are necessary.

$43.68 \to 4_\wedge 3.68$ or 4.368×10^1

Write 43.68 in scientific notation.

$4.368 \times 10^1 \times 10^{-5}$.

Multiply powers of ten.

$$4.368 \times 10^1 \times 10^{-5} = 4.368 \times 10^{1-5}$$
$$= 4.368 \times 10^{-4}$$

How to

Divide numbers in scientific notation:

1. Divide the first factors using the rules of signed numbers.

2. Divide the power-of-10 factors using the laws of exponents.
3. Examine the first factor of the quotient (Step 1) to see if its value is equal to 1 or between 1 and 10.
  a. If so, write the results of Steps 1 and 2.
  b. If not, shift the decimal so that the first factor is equal to 1 or is between 1 and 10, and adjust the exponent of the power-of-10 factor accordingly.

EXAMPLE To divide:

(a) $\dfrac{3 \times 10^5}{2 \times 10^2}$ (b) $\dfrac{1.44 \times 10^{-3}}{6 \times 10^{-5}}$

(a) $\dfrac{3 \times 10^5}{2 \times 10^2} = \dfrac{3}{2} \times 10^{5-2}$

The first factor is usually written in decimal notation.

$= 1.5 \times 10^3$

Because 1.5 is between 1 and 10, no adjustments are necessary.

(b) $\dfrac{1.44 \times 10^{-3}}{6 \times 10^{-5}} = \dfrac{1.44}{6} \times 10^{-3-(-5)}$

$= 0.24 \times 10^{-3+5}$

$= 0.24 \times 10^2$

0.24 is less than 1, so adjustments are necessary.

$= 2.4 \times 10^{-1} \times 10^2$

Shift decimal and adjust power of 10.

$= 2.4 \times 10^{-1} \times 10^2$

Multiply powers of 10.

$= 2.4 \times 10^1.$

▲ **APPLICATION: Distance between Star and Earth: Using Scientific Notation for Very Large or Very Small Numbers**

How to

Solve problems with measures written in scientific notation:

1. Express the unknown as a variable.

> 2. Express the known and unknown information in a proportion relating the quantities.

EXAMPLE A star is 4.2 light-years from Earth. If 1 light-year is 5.87×10^{12} miles, to find the number of miles from Earth the star is:

Pair 1: 1 light-year = 5.87×10^{12} mi

Pair 2: 4.2 light-years = x mi

Express the given information in a direct proportion of two fractions relating light-years to miles.

Estimation   The star will be more than 5.87×10^{12} miles away.

$$\frac{1 \text{ light-year}}{4.2 \text{ light-years}} = \frac{5.87 \times 10^{12} \text{ mi}}{x \text{ mi}}$$

Pair 1 becomes the two numerators.
Pair 2 becomes the two denominators.
(Direct proportion)

$$\frac{1}{4.2} = \frac{5.87 \times 10^{12}}{x}$$

Cross multiply.

$$x = (\,4.2\,)(\,5.87 \times 10^{12}\,)$$

4.2 or 4.2×10^0

$$x = 24.654 \times 10^{12}$$

Perform scientific notation adjustment.

$$x = 2.4654 \times 10^{12+1}$$

$N \rightarrow O = +1$

$$x = 2.4654 \times 10^{13}$$

or 2.5×10^{13}

Rounded first factor.

Interpretation   The star is 2.5×10^{13} miles from Earth.

⌞

■ 12–4 Polynomials

A **polynomial** is an algebraic expression in which the exponents of the variables are nonnegative integers.

A **monomial** is a polynomial containing one term. A term may have more than one factor.

$3, \quad -2x, \quad 5ab, \quad 7xy^2, \quad \dfrac{3a^2}{4}$ are monomials.

A **binomial** is a polynomial containing two terms.

$x + 3, \quad 2x^2 - 5x, \quad x + \dfrac{y}{4}, \quad 3(x - 1) + 2$ are binomials.

A **trinomial** is a polynomial containing three terms.

$a + b + c, \quad x^2 - 3x + 4, \quad x + \dfrac{2a}{7} - 5$ are trinomials.

How to

Identify polynomials, monomials, binomials, and trinomials:

1. Write all variables in the numerator if necessary.
2. Exclude expressions that have a negative exponent in any term.
3. Identify the expression based on the number of terms it contains.

EXAMPLE   To identify the expressions as a polynomial, monomial, binomial, trinomial, or none of these:

(a) $x^2y - 1$ (b) $4(x - 2)$

(c) $\dfrac{2x + 5}{2y}$ (d) $3x^3 - x^2 + 3x - 5$

(a) $x^2y - 1$	Binomial (2 terms).
(b) $4(x - 2)$	Monomial (1 term).
(c) $\dfrac{2x + 5}{2y}$	This one-term expression is **not a monomial** because **it is not a polynomial** (the exponent of y would be -1 if it were in the numerator).
(d) $3x^3 - x^2 + 3x - 5$	Polynomial (4 terms).

The **degree of a term** that has only one variable with a nonnegative exponent is the same as the exponent of the variable. The **degree of a constant** is zero. The degree of a term with more than one variable is the sum of the exponents provided each variable has a positive exponent.

How to

Identify the degree of a term:

1. Exponents of variable factors must be integers greater than zero.
2. For a term that is a constant, the degree is zero.

3. For a term that has only one variable factor, the degree is the same as the exponent of the variable.
4. For a term that has more than one variable factor, the degree is the sum of the exponents of the variables.

EXAMPLE   To identify the degree of each term in the polynomial $5x^3y^2 + 2x^2 - 3x + 3$:

$5x^3y^2$ Exponents of variables are positive integers; $3 + 2 = 5$. Degree is 5.

$2x^2$ Exponent of variable is positive. Degree is 2.

$-3x$ Exponent of variable is 1. Degree is 1.

3 This is a constant term. Degree is 0.

A **constant term** has degree 0. A **linear term** has degree 1. A **quadratic term** has degree 2. A **cubic term** has degree 3.

The **degree of a polynomial** that has only one variable and only positive integral exponents is the degree of the term with the largest exponent.

A **linear polynomial** has degree 1. A **quadratic polynomial** has degree 2. A **cubic polynomial** has degree 3.

How to

Identify the degree of a polynomial:

1. Identify the degree of each term of the polynomial.
2. Compare the degrees of each term of the polynomial and select the greatest degree as the degree of the polynomial.

EXAMPLE   To identify the degree of the polynomial $5x^3y^2 + 3x^2y^2 + 4xy^3 + 8$:

$5x^3y^2$ The first term has degree $3 + 2$ or 5.

$3x^2y^2$ The degree of this term is $2 + 2$ or 4.

$4xy^3$ The degree of this term is $1 + 3$ or 4.

8 The degree of this constant term is zero.

$5, 4, 4, 0$ Compare the degrees of each of the terms of the polynomial.

5 The greatest degree is 5.

The **degree of the polynomial** $5x^3y^2 + 3x^2y^2 + 4xy^3 + 8$ is 5.

A degree of the polynomial is arranged in **descending order** of a variable if it is arranged with the term that has the highest degree of the variable listed first and terms with smaller degrees of the variable listed in decreasing order by degree. Polynomials are arranged in **ascending order** if they have the term that has the lowest degree listed first and subsequent terms listed in order of the increasing degrees.

TIP

Most common arrangement of polynomials

Polynomials are most often arranged in **descending order** so the degree of the polynomial is the degree of the first term.

How to

Arrange polynomials in descending order of a variable:

1. Identify the variable on which the terms of the polynomial will be arranged if the polynomial has more than one variable.
2. Compare the degrees of the selected variable for each term.
3. List the term with highest degree of the specified variable first.
4. Continue to list the terms of the polynomial in descending order of the selected variable.

EXAMPLE   To arrange the polynomial $5x^2y^3 + 4x^4y^2 - 7xy^5 + 3x^3y - 8$ in descending powers of x:

$5\,x^2\,y^3 + 4\,x^4\,y^2 - 7\,x\,y^5 + 3\,x^3\,y - 8$

Identify the variable on which the polynomial will be arranged. Examine the degree of x in each term.

Comparing the degrees of x for each term shows $4x^4y^2$ to be the greatest degree of x.

$4x^4y^2 + 3x^3y + 5x^2y^3 - 7xy^5 - 8$

The terms are arranged in descending powers (degrees) of x. The degrees of y are not considered.

CHAPTER 13

Products and Factors

- **13–1 The Distributive Principle and Common Factors**
- **13–2 Multiplying Polynomials**
- **13–3 Factoring Special Products**
- **13–4 Factoring General Trinomials**

■ 13–1 The Distributive Principle and Common Factors

An algebraic expression written as the sum or difference of terms is written in **expanded form**. **Factoring** is the process of writing an expression as multiplication. A factor that appears in all the terms of an expression is a **common factor** of the expression.

How to

Factor an expression containing a common factor:

1. Identify the *greatest* factor common to *each* term of the expression.
2. Divide each term by the common factor.
3. Rewrite the expression as the indicated product of the greatest common factor (GCF) and the quotients in Step 2.

EXAMPLE   To factor $10a^2 + 6a$ completely:

$$10a^2 + 6a = 2a \left(\frac{10a^2}{2a} + \frac{6a}{2a} \right)$$ Because 2 and a are common factors, the greatest common factor is $2a$. Divide each term by $2a$.

$$10a^2 + 6a = 2a(5a + 3)$$ Factored form.

TIP

When is it necessary to write a 1?

It is sometimes necessary to write the number 1. When is this the case? When 1 is a *factor* (multiplicative property of 1, $1 \cdot n = n$), the 1 does not have to be written. When the exponent of a factor is 1 (definition of the exponent of 1, $a^1 = a$), the 1 does not have to be written. **When 1 is a term** instead of a factor **it must be written.**

EXAMPLE   To factor $2x^2 + 4x^3$ completely:

$$2x^2 + 4x^3 = 2x^2 \left(\frac{2x^2}{2x^2} + \frac{4x^3}{2x^2} \right)$$ Because 2 and x^2 are common factors, the greatest common factor is $2x^2$.

$$= 2x^2(1 + 2x)$$ Factored form.

● **RELATED TOPIC:** Dividing algebraic expressions

■ 13–2 Multiplying Polynomials

A form of long multiplication can be used to multiply polynomials. The process requires that each term of one polynomial must be multiplied by each term of the second polynomial. Repeated application of the distributive property is necessary.

EXAMPLE   To multiply $(2x + 3)(3x - 1)$:

$(2x + 3)(3x - 1) = 2x (3x - 1) + 3 (3x - 1)$ Apply the distributive property.

$= 6x^2 - 2x + 9x - 3$ Combine like terms.

$= 6x^2 + 7x - 3$

A shortcut that guides us through this process is called the **FOIL method** for multiplying two binomials.

In this systematic process using the acronym FOIL, F represents the product of the *first* term in each factor, O represents the product of the two *outer* terms, I represents the product of the *inner* two terms, and L represents the product of the *last* term in each factor. If the inner and outer products are like terms, they should be combined.

Products Factors

How to

Multiply two binomials using the FOIL method:

1. Write the product of the *first* term of each factor.
2. Write the product of the two *outer* terms of the factors.
3. Write the product of the two *inner* terms of the factors.
4. Write the product of the *last* term of each factor.
5. Combine like terms.

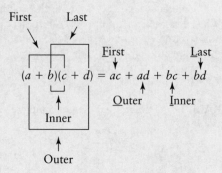

$$(a + b)(c + d) = ac + ad + bc + bd$$

EXAMPLE   To use the FOIL method to multiply $(2x - 3)(x + 1)$:

$$(2x - 3)(x + 1) = 2x^2 + 2x - 3x - 3 \qquad \text{Combine like terms.}$$
$$= 2x^2 - x - 3$$

The product of a binomial and trinomial or two trinomials extends the process of repeatedly applying the distributive property.

EXAMPLE   To multiply $(x + 5)(x^2 + 3x + 1)$:

$(x + 5)(x^2 + 3x + 1) =$ Multiply each term of the binomial times the entire trinomial.

$x(x^2 + 3x + 1) + 5(x^2 + 3x + 1) =$ Distribute.

$x^3 + 3x^2 + x + 5x^2 + 15x + 5 =$ Arrange terms in descending order.

$x^3 + 3x^2 + 5x^2 + x + 15x + 5 =$ Combine like terms.

$x^3 + 8x^2 + 16x + 5$

All polynomials can be multiplied using the process in the previous example. However, there are some special expressions where shortcut processes can be used. When the polynomials match a specific pattern, the product also matches a specific pattern. We call the special expressions **special products.**

Pairs of factors like $(a + b)(a - b)$ are called the **sum and difference of the same two terms** or **conjugate pairs.** The product, $a^2 - b^2$, is called the **difference of two perfect squares.** The outer and inner products in the FOIL method add to zero.

How to

Mentally multiply the sum and difference of the same two terms:

1. Square the first term of the first factor.
2. Insert a minus symbol.
3. Square the second term of the first factor.

Symbolically, $(a + b)(a - b) = a^2 - b^2$

EXAMPLE   To find the products mentally:

(a) $(x + 2)(x - 2)$

(b) $(2m + 3)(2m - 3)$

(a) $(x + \boxed{2})(x - \boxed{2}) =$ Square x; insert a minus symbol; square 2.

$x^2 - \boxed{4}$

(b) $(2m + \boxed{3})(2m - \boxed{3}) =$ $(2m)^2 = 4m^2$; insert minus symbol; $3^2 = 9$.

$4m^2 - \boxed{9}$

Pairs of factors like $(a + b)(a + b)$ can be written in exponential form as $(a + b)^2$. An expression written as the product of the *same* two binomials or as the **square of a binomial** is called a **binomial square.** The expanded trinomial is called a **perfect square trinomial.**

How to

Square a binomial:

1. Square the first term.

2. Multiply the two terms, then double the product for the middle term of the product.
3. Square the last term.

Symbolically, $(a + b)^2 = a^2 + 2ab + b^2$

$$(a - b)^2 = a^2 - 2ab + b^2$$

EXAMPLE To square the binomials:

   (a) $(2x + 5)^2$ (b) $(3a - 2b)^2$

   (a) $(2x + 5)^2$

$(2x)^2 = 4x^2$	Square the first term.
$2\left[\,2x(+5)\,\right] = 2\left[\,+10x\,\right] = 20x$	Multiply the two terms, then double.
$(+5)^2 = 25$	Square the last term.
$(2x + 5)^2 = 4x^2 + 20x + 25$	Write as a perfect square trinomial.

   (b) $(3a - 2b)^2$

$(3a)^2 = 9a^2$	Square the first term.
$2\left[\,(3a)(-2b)\,\right] = 2\,(-6ab) = -12ab$	Multiply the two terms, then double.
$(-2b)^2 = 4b^2$	Square the last term.
$(3a - 2b)^2 = 9a^2 - 12ab + 4b^2$	Write as a perfect square trinomial.

└

    Other special products have terms that add to zero and allow a shortcut process.

How to

Multiply a binomial and a trinomial of the types
$(a + b)(a^2 - ab + b^2)$ and $(a - b)(a^2 + ab + b^2)$:

1. Cube the first term of the binomial.
2. If the binomial is a sum, insert a plus sign. If it is a difference, insert a minus sign.
3. Cube the second term of the binomial.

Symbolically, $(a + b)(a^2 - ab + b^2) = a^3 + b^3$

$$(a - b)(a^2 + ab + b^2) = a^3 - b^3$$

TIP

Using symbolic patterns

The pattern $(a + b)(a^2 - ab + b^2) = a^3 + b^3$ is wordy to describe in sentence form. The pattern is more easily identified when it is written symbolically.

When this pattern is applied to different polynomials, we substitute for the letters a and b in the pattern.

Multiply $(2x + 3)(4x^2 - 6x + 9)$.

Do these factors match the pattern? Let a represent $2x$ and b represent 3.

$a^2 = (2x)^2 = 4x^2$ Matches pattern.

$b^2 = 3^2 = 9$ Matches pattern.

$ab = (2x)(3) = 6x$ Matches pattern.

Sign of binomial is positive.

Middle sign of trinomial is negative. } Matches pattern.

Apply the pattern of the product.

$(a + b)(a^2 - ab + b^2) = a^3 + b^3$

$a^3 = (2x)^3 = 8x^3$ $b^3 = 3^3 = 27$

Then, $(2x + 3)(4x^2 - 6x + 9) = 8x^3 + 27$.

EXAMPLE   To multiply $(4 - d)(16 + 4d + d^2)$:
Do the factors match a pattern? Yes. Here, a represents 4 and b represents d. Cube the 4 and the d; insert a minus sign between the cubes because the binomial is a difference.

$(4 - d)(16 + 4d + d^2) =$ **64** $- d^3$ $4^3 = 64$

■ 13–3 Factoring Special Products

The key to writing special products in factored form is to *recognize* the product as a special product.

How to

Factor the difference of two perfect squares:

1. Recognize the binomial as the difference of two perfect squares.
2. Take the square root of the first term.
3. Take the square root of the second term.
4. Write one factor as the *sum* of the square roots found in Steps 2 and 3, and write the other as the *difference* of the square roots found in Steps 2 and 3.

Symbolically, $a^2 - b^2 = (a + b)(a - b)$

EXAMPLE   To factor the special products, which are the differences of two perfect squares:

   (a) $a^2 - 9$ (b) $16m^2 - 49$

   (a) $a^2 - 9$ First and last terms are perfect squares. Terms are expressed as a difference.

$$\sqrt{a^2} = \boxed{a}$$ Take the square root of the first term.

$$\sqrt{9} = \boxed{3}$$ Take the square root of the second term.

$$(a + \boxed{3})(a - \boxed{3})$$ One factor is the *sum* of the square roots and one factor is the *difference* of the square roots.

   (b) $16m^2 - 49$ First and last terms are perfect squares. Terms are expressed as a difference.

$$\sqrt{16m^2} = \boxed{4m}$$ Take the square root of the first term.

$$\sqrt{49} = \boxed{7}$$ Take the square root of the second term.

$$(\boxed{4m} + \boxed{7})(\boxed{4m} - \boxed{7})$$ One factor is the *sum* of the square roots and one factor is the *difference* of the square roots.

How to

Factor a perfect square trinomial:

1. Recognize the trinomial as a perfect square trinomial.
    a. The first term is a perfect square.
    b. The third term is a perfect square.
    c. The middle term is twice the product of the square roots of the first and third terms.

2. Write the square root of the first term.
3. Write the sign of the middle term.
4. Write the square root of the last term.
5. Indicate the square of this binomial quantity.

Symbolically,

$$a^2 + 2ab + b^2 = (a + b)^2$$

$$a^2 - 2ab + b^2 = (a - b)^2$$

EXAMPLE   To factor the perfect-square trinomial $4m^2 - 12m + 9$:

$4m^2 - 12m + 9$	The product matches the pattern.
$4m^2 = (2m)^2$	First term is a perfect square.
$9 = (3)^2$	Third term is a perfect square.
$-12m = 2[(2m)(-3)]$	Middle term is twice the product of the square roots of the first and third terms.
$(2m - 3)^2$	Square root of first term ($2m$); square root of last term (3); sign of middle term ($-$); square the binomial.

∟

The sum or difference of two **perfect cubes,** such as $x^3 + y^3$ or $a^3 - b^3$ is a special product of a binomial and a trinomial.

How to

Factor the sum of two perfect cubes:

1. Identify the binomial as the sum of two perfect cubes.
2. Write the binomial factor as the *sum* of the cube roots of the two terms.
3. Write the trinomial factor as the square of the first term from Step 2, *minus* the product of the two terms from Step 2, plus the square of the second term from Step 2.

Symbolically,

$$a^3 + b^3 = (a + b)(a^2 - ab + b^2)$$

Products
Factors

EXAMPLE To factor $8x^3 + 27$:

$8x^3 + 27$ — Both terms are perfect cubes and the binomial is a sum.

$\sqrt[3]{8x^3} = \boxed{2x}\,;\ \sqrt[3]{27} = \boxed{3}$ — Find the cube root of each term.

$\left(\boxed{2x} + \boxed{3}\right)$ — The binomial factor.

$\left[\left(\boxed{2x}\right)^2 - \boxed{2x}\left(\boxed{3}\right) + \boxed{3}^2\right]$ — The trinomial factor.

$(2x + 3)(4x^2 - 6x + 9)$ — The factors of the binomial.

How to

Factor the difference of two perfect cubes:

1. Identify the binomial as the difference of two perfect cubes.
2. Write the binomial factor as the _difference_ of the cube roots of the two terms.
3. Write the trinomial factor as the square of the first term from Step 2, _plus_ the product of the two terms from Step 2, plus the square of the second term from Step 2.

Symbolically,

$$a^3 - b^3 = (a - b)(a^2 + ab + b^2)$$

EXAMPLE To factor $64x^3 - 1$:

$64x^3 - 1$ — Both terms are perfect cubes and the binomial is a difference.

$\sqrt[3]{64x^3} = \boxed{4x}\,;\ \sqrt[3]{1} = \boxed{1}$ — Find the cube root of each term.

$\left(\boxed{4x} - \boxed{1}\right)$ — The binomial factor.

$\left[\left(\boxed{4x}\right)^2 + \boxed{4x}\ (1) + \boxed{1}^2\right]$ — The trinomial factor.

$(4x - 1)(16x^2 + 4x + 1)$ — The factors of the binomial.

■ 13–4 Factoring General Trinomials

Many trinomials do not have a common factor and do not match the pattern of a special product. Some will still factor as the product of two binomials.

How to

Factor a trinomial with a squared term that has a coefficient of 1:

1. Ensure the trinomial is arranged in descending powers of the variable.
2. Write the factor pairs of the third term, taking into consideration the sign of the third term.
3. Select the factor pair that adds to the coefficient of the middle term of the trinomial.
4. Write the binomial factors of the trinomial with the variable as the first term of each factor and the two factors from Step 3 as the second terms of the binomials.

EXAMPLE   To factor the trinomial $x^2 - 4x - 12$:

$x^2 - 4x - 12$ The terms are arranged in descending powers of x.

$1 \cdot -12$ Write the factor pairs of the third term, taking into consideration the sign of the third term.

$2 \cdot -6$

$3 \cdot -4$

$2 + (-6) = -4$ Select the factor pair that adds to the coefficient of the middle term of the trinomial.

$(x + 2)(x - 6)$ Write the binomial factors of the trinomial using the two factors from Step 3.

● RELATED TOPIC: Factor pairs

Algebraic expressions that have no factors common to every term in the expression may have common factors for some groups of terms. In these cases, the expression is written in factored form. In the expression $2x^2 - 2xb + ax - ab$, there are four terms and no factor is common to every term. To factor, write the expression as two terms by *grouping* the first two terms and the last two terms; then look for common factors in each grouping.

$$(2x^2 - 2xb) + (ax - ab)$$

The first grouping, $(2x^2 - 2xb)$, has a common factor of $2x$ and can be written in factored form as $2x(x - b)$. The second grouping, $(ax - ab)$, has a common factor of a and can be written in factored form as $a(x - b)$. If we look at the entire expression $2x(x - b) + a(x - b)$, we see the common binomial factor $(x - b)$. When we factor this common factor from both terms, we have $(x - b)(2x + a)$. The expression is now one term that is written as a product of two factors. We can check the factoring by using the distributive property or the FOIL method of multiplying.

$$(x - b)(2x + a) = 2x^2 + ax - 2xb - ab$$

This result is the same as the original expression with the terms in a different order.

TIP

Groupings are not necessarily factors

In the expression $(2x^2 - 2xb) + (ax - ab)$, the groupings $(2x^2 - 2xb)$ and $(ax - ab)$ are not factors. Notice that they are groupings that are added. Additional factoring techniques are used to write the expression in factored form.

How to

Factor a general trinomial of the form $ax^2 + bx + c$ by grouping:

1. Multiply the coefficient of the first term by the coefficient of the last term.
2. Factor the product from Step 1 into a pair of factors:
   a. whose *sum* is the coefficient of the middle term if the sign of the last term is positive, or
   b. whose *difference* is the coefficient of the middle term if the sign of the last term is negative.
3. Rewrite the trinomial so that the middle term is the sum or difference from Step 2.
4. Group the polynomial with four terms from Step 3 into two groups of two terms each.
5. Factor out the common factor from each grouping.

EXAMPLE To factor $6x^2 + 19x + 10$ by grouping:

$6(10) = 60$ Multiply the coefficients of the first and third terms.

$60 = (1)60$ List all factor pairs of 60.

(2)30

(3)20

(4)15   Identify the pair that *adds* to 19, since
                              the sign of the last term is positive.

(5)12

(6)10

$6x^2 + 19x + 10 =$ Separate $+19x$ into two terms using the
                              coefficients 4 and 15.

$6x^2 + 4x + 15x + 10 =$ Group into two groupings of two terms
                              each.

$(6x^2 + 4x) + (15x + 10) =$ Factor the common factors from each
                              grouping.

$2x(3x + 2) + 5(3x + 2) =$ Factor the common binomial.

$(3x + 2)(2x + 5)$ Check using the FOIL method.

How to

Factor any binomial or trinomial:

Perform the following steps in order.

1. Ensure the trinomial is arranged in descending powers of the variable.
2. Factor the greatest common factor (if any).
3. Check the binomial or trinomial to see if it is a special product.
   a. If it is the difference of two perfect squares, use the pattern.
   b. If it is a perfect-square trinomial, use the pattern.
   c. If it is the sum or difference of two perfect cubes, use the pattern.
4. If there is no special product, factor as a general trinomial.
5. Examine each factor to see that it cannot be factored further.
6. Check final factoring by multiplying.

**Products
Factors**

EXAMPLE   To completely factor $4x^3 - 2x^2 - 6x$:

$2x(2x^2 - x - 3)$ $2x$ is the common factor.

$2x(2x - 3)(x + 1)$ Factor the trinomial but *keep* the $2x$ factor. Check
                            by multiplying.

Shortening the process for factoring by grouping

The following is a variation of the factor-by-grouping method. The variation is mathematically sound and employs a strategy using the property of 1 that is often overlooked.

Factor $6x^2 + 7x - 20$.

$6x^2 + 7x - 20$	Multiply the coefficients of the first and third terms: $6(-20) = -120$.

$120 =$

1	120
2	60
3	40
4	30
5	24
6	20
8	15
10	12

A negative product means the factor pair has unlike signs. List all factor pairs of 120.

Identify the factor pair with a *difference* of +7. The larger factor will be positive, the sign of the middle term.

$-8 + 15 = 7$

Variation in procedure begins here.

$$\dfrac{(6x\quad)(6x\quad)}{6}$$

Use the coefficient of the first term as the coefficient of the first term in *each* binomial. This gives you an extra factor of 6. Compensate by *dividing* the expression by 6.

$$\dfrac{(6x - 8)(6x + 15)}{6} =$$

Use the factors of -120 that have an algebraic sum of 7 as the second term of each binomial.

$$\dfrac{2(3x - 4)(3)(2x + 5)}{6} =$$

Factor the common factors from each binomial.

$$\dfrac{6(3x - 4)(2x + 5)}{6} =$$

The factors of the trinomial will be the two binomial factors. The factor 6 reduces.

$$(3x - 4)(2x + 5)$$

Check using the FOIL method.

Why does this variation work? The coefficient of the first term in the original trinomial, 6, was used in each binomial. Thus, the resulting product would have an extra factor of 6. This extra factor of 6 must be removed to obtain the correct factors. **Now, let's shorten the process again.**

(continued)

Once you have selected the pair of factors whose sum or difference matches the middle term, *make two fractions using the factor pair as the denominators*. The *leading coefficient of the trinomial will be the numerator of each fraction*.

$6x^2 + 7x - 20$ $6(-20) = -120$

The factor pair of -120 that has a sum of $+7$ is -8 and $+15$. Make fractions and reduce each fraction.

$$\frac{6}{-8} = \frac{3}{-4} \qquad \frac{6}{+15} = \frac{2}{+5}$$

Each fraction gives the coefficients of one of the binomial factors. 3, -4 and 2, $+5$.

$(3x - 4)(2x + 5)$

If the trinomial has a common factor, you must factor the common factor before finding the factor pair and writing the fractions. The common factor will be part of the final answer.

Factor $30x^2 + 8 - 32x$.

$30x^2 - 32x + 8$ **Arrange in descending powers of x.**

$2(15x^2 - 16x + 4)$ **Factor any common factors.**

$15(4) = 60$ **Factor pairs of 60.**

1	60
2	30
3	20
4	15
5	12
6	10

Identify the pair that has a *sum* of -16.
$-6 + (-10) = -16$

Make fractions of the factor pair -6 and -10 and the leading coefficient of the trinomial, 15. Reduce each fraction.

$$\frac{15}{-6} = \frac{5}{-2} \qquad \frac{15}{-10} = \frac{3}{-2}$$

Write the factors using the coefficients 5, -2 and 3, -2. Don't forget the common factor.

$2(5x - 2)(3x - 2)$

The steps for factoring a binomial or trinomial of degree 2 are presented visually in the **factoring flow chart.**

Products Factors

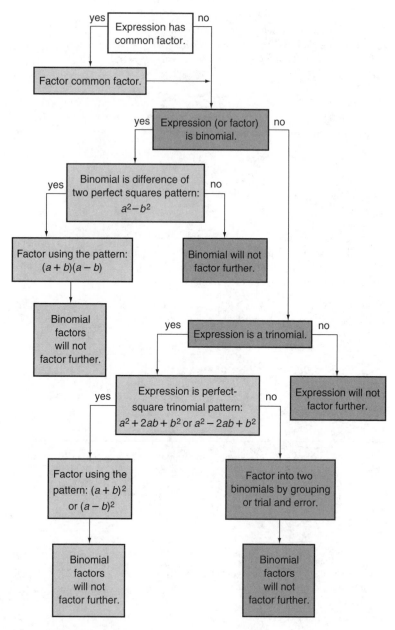

Factoring flow chart for binomials and trinomials of degree two.

CHAPTER 14

Roots and Radicals

- **14–1 Roots and Notation Conventions**
- **14–2 Simplifying Square-Root Expressions**
- **14–3 Basic Operations with Square-Root Radicals**
- **14–4 Equations with Squares and Square Roots**

■ 14–1 Roots and Notation Conventions

There are two common notation conventions for roots: **rational exponent notation** and **radical notation.**

How to

Convert between rational exponent notation and radical notation:

Rational exponent to radical:

1. Write the numerator of a rational exponent as the power.
2. Write the denominator of the rational exponent as the index of the root.

Symbolically, $x^{\text{power/root}} = \sqrt[\text{root}]{x^{\text{power}}}$ or $\left(\sqrt[\text{root}]{x}\right)^{\text{power}}$

Radical to rational exponent:

1. Write the power as the numerator of the rational exponent.
2. Write the index of the root as the denominator of the rational exponent.

Symbolically, $\sqrt[\text{root}]{x^{\text{power}}} = x^{\text{power/root}}$

221

EXAMPLE To write $x^{2/3}$ in radical form:

$x^{2/3}$ Write the numerator of the rational exponent as the power.
Write the denominator of the rational exponent as the index of
the root.

└ $\sqrt[3]{x^2}$

EXAMPLE To write $\sqrt[4]{x^3}$ in rational exponent notation:

$\sqrt[4]{x^3}$ Write the power of the radicand as the numerator of the
fractional exponent. Write the index as the denominator of the
fractional exponent.

$x^{3/4}$
└

■ 14–2 Simplifying Square-Root Expressions

How to

*Simplify radicals using rational exponents and the laws
of exponents:*

1. Convert the radicals to equivalent expressions using rational
   exponents.
2. Apply the laws of exponents and the arithmetic of fractions.
3. Convert simplified expressions back to radical notation if desired.

EXAMPLE To convert the radical expressions to equivalent expressions
using rational exponents and simplify if appropriate.

(a) $\sqrt[3]{x}$ (b) $\sqrt[5]{2y}$ (c) $(\sqrt{ab})^3$ (d) $\sqrt[4]{16b^8}$

(e) $(\sqrt[3]{27xy^5})^4$

(a) $\sqrt[3]{x} = x^{1/3}$ (b) $\sqrt[5]{2y} = (2y)^{1/5}$ or $2^{1/5}y^{1/5}$

(c) $(\sqrt{ab})^3 = (ab)^{3/2}$ or $a^{3/2}b^{3/2}$

(d) $\sqrt[4]{16b^8} = (2^4b^8)^{1/4} = (2^4)^{1/4}(b^8)^{1/4} = 2b^2$

16 is a perfect fourth power; $2^4 = 16$. Therefore, $(2^4)^{1/4} = 2$.

(e) $(\sqrt[3]{27xy^5})^4 = (3^3x^1y^5)^{4/3} = (3^3)^{4/3}x^{4/3}(y^5)^{4/3} = 3^4x^{4/3}y^{20/3}$

$= 81x^{4/3}y^{20/3}$

27 is a perfect cube; $3^3 = 27$. Therefore, $27^{4/3} = (3^3)^{4/3} = (3)^4 = 81$.
└

How to

Simplify expressions with fractional exponents:

1. Apply the same steps used to simplify integral exponents.
2. Reduce fractional exponents if necessary.

● **RELATED TOPIC: Laws of exponents**

EXAMPLE   To perform the following operations and simplify and express answers with positive exponents in lowest terms:

(a) $(x^{3/2})(x^{1/2})$ (b) $(3a^{1/2}b^3)^2$ (c) $\dfrac{x^{1/2}}{x^{1/3}}$ (d) $\dfrac{10a^3}{2a^{1/2}}$

(a) $(x^{3/2})(x^{1/2}) = x^{3/2+1/2} = x^{4/2} = x^2$ Add exponents.

(b) $(3a^{1/2}b^3)^2 = 3^2ab^6 = 9ab^6$ Multiply exponents,
$\dfrac{1}{2} \cdot 2 = 1; \, 3 \cdot 2 = 6$

(c) $\dfrac{x^{1/2}}{x^{1/3}} = x^{1/2-1/3} = x^{1/6}$ Subtract exponents.
$\dfrac{1}{2} - \dfrac{1}{3} = \dfrac{3}{6} - \dfrac{2}{6} = \dfrac{1}{6}$

(d) $\dfrac{10a^3}{2a^{1/2}} = 5a^{3-1/2} = 5a^{5/2}$ Reduce coefficients. Subtract exponents.
$3 - \dfrac{1}{2} = \dfrac{3}{1} - \dfrac{1}{2} = \dfrac{6}{2} - \dfrac{1}{2} = \dfrac{5}{2}$

● **RELATED TOPIC: Adding, subtracting, and multiplying fractions**

To **factor an algebraic expression** is to write it as the indicated product of two or more factors, that is, as a multiplication.

How to

Simplify square-root radicals containing perfect-square factors:

1. If the radicand is a perfect square, express it as a square root without the radical sign.

Roots and Radicals

2. If the radicand is *not* a perfect square, factor the radicand into as many perfect-square factors as possible. The square roots of the perfect-square factors appear *outside* the radical, and the other factors stay *inside* (under) the radical sign.

If the radicand is *not* a perfect square and *cannot* be factored into one or more perfect-square factors, it is already in simplified form.

EXAMPLE   To simplify the radicals:

(a) $\sqrt{32}$ (b) $\sqrt{y^7}$ (c) $\sqrt{18x^5}$ (d) $\sqrt{75xy^3z^5}$ (e) $\sqrt{7x}$

When factoring coefficients, some perfect squares that can be used are 4, 9, 16, 25, 36, 49, 64, and 81.

(a) $\sqrt{32} = \sqrt{16 \cdot 2}$ What is the *largest* perfect-square factor of 32? 4 is a factor of 32, but 16
$= \sqrt{16}\left(\sqrt{2}\right) = 4\sqrt{2}$ is also a factor of 32. Use the *largest* perfect-square factor.

(b) $\sqrt{y^7} = \sqrt{y^6 \cdot y^1} = \sqrt{y^6}\left(\sqrt{y}\right)$ The largest perfect-square factor of y^7 is y^{7-1}, or y^6.
$= y^3\sqrt{y}$

(c) $\sqrt{18x^5}$ Factor radicand into as many perfect square factors as you can.

$= \sqrt{9 \cdot 2 \cdot x^4 \cdot x^1}$ Write each factor as a separate radical.

$= \sqrt{9}\left(\sqrt{2}\right)\left(\sqrt{x^4}\right)\left(\sqrt{x}\right)$ Take the square root of the two perfect squares.

$= 3\left(\sqrt{2}\right)\left(x^2\right)\sqrt{x}$ Multiply coefficients. Multiply radicals.

$= 3x^2\sqrt{2x}$

In the previous example we showed each step in the simplifying process. However, we customarily do most of these steps mentally.

(d) $\sqrt{75xy^3z^5} = \sqrt{25 \cdot 3 \cdot x \cdot y^2 \cdot y \cdot z^4 \cdot z}$
$= 5yz^2\sqrt{3xyz}$

Write the square roots of the perfect-square factors outside the radical sign. The other factors are written under the radical sign.

(e) $\sqrt{7x} = \sqrt{7x}$ 7 and x contain no perfect-square factors.

● **RELATED TOPIC: Square roots**

Finding perfect-square factors

- Whole-number perfect squares: 1, 4, 9, 16, 25, 36, 49, 64, 81, 100, 121, 144, ...
- 1 is a factor of any number: $8 = 8 \cdot 1$. To factor using the perfect square 1 does not simplify a radicand.
- Any variable with an exponent higher than 1 is a perfect square or has a perfect-square factor. We assume variables represent positive values.
- Perfect-square variables have even-numbered exponents.

$x^2, x^4, x^6, x^8, x^{10}, \ldots$

- Variables with a perfect-square factor:

$x^3 = x^2 \cdot x^1, \qquad x^5 = x^4 \cdot x^1$
$x^7 = x^6 \cdot x^1, \qquad x^9 = x^8 \cdot x^1$

- A convenient way to keep track of perfect-square factors is to circle them. The square roots of circled factors are written outside the radical sign. The uncircled factors stay in the radicand as is.

$$\sqrt{75ab^4c^3} = \sqrt{\textcircled{25} \cdot 3 \cdot a^1 \cdot \textcircled{b^4} \cdot \textcircled{c^2} \cdot c^1}$$
$$= 5b^2c\sqrt{3ac}$$

■ 14–3 Basic Operations with Square-Root Radicals

When the radicands are identical and the radicals have the same order or index, the radicals are **like radicals.**

How to

Add or subtract like square-root radicals:

1. Add or subtract the coefficients of the radicals.
2. Use the common radicand as a factor in the solution.

Symbolically, $a\sqrt{b} + c\sqrt{b} = (a + c)\sqrt{b}$

Roots and Radicals

EXAMPLE   To add or subtract radical expressions:

(a) $3\sqrt{7} + 2\sqrt{7}$ (b) $4\sqrt{2} + \sqrt{2}$

(c) $5\sqrt{3} + 7\sqrt{5} + 2\sqrt{3} - 4\sqrt{5}$

(a) $3\sqrt{7} + 2\sqrt{7} = 5\sqrt{7}$

Radicals are like, so add coefficients.

(b) $4\sqrt{2} + \sqrt{2} = 5\sqrt{2}$

When no coefficient is written in front of a radical, the coefficient is 1.

(c) $5\sqrt{3} + 7\sqrt{5} + 2\sqrt{3} - 4\sqrt{5} = 7\sqrt{3} + 3\sqrt{5}$

Combine like radical terms.

● **RELATED TOPIC: Adding like terms**

How to

Multiply square-root radicals:

1. Multiply coefficients to give the coefficient of the product.
2. Multiply radicands to give the radicand of the product.
3. Simplify if possible.

Symbolically, $a\sqrt{b} \cdot c\sqrt{d} = ac\sqrt{bd}$

EXAMPLE   To multiply the following radicals:

(a) $\sqrt{3} \cdot \sqrt{5}$ (b) $\sqrt{\dfrac{7}{8}} \cdot \sqrt{\dfrac{2}{3}}$ (c) $3\sqrt{2} \cdot 4\sqrt{3}$

(a) $\sqrt{3} \cdot \sqrt{5} = \sqrt{15}$ Multiply radicands.

(b) $\sqrt{\dfrac{7}{8}} \cdot \sqrt{\dfrac{2}{3}} = \sqrt{\dfrac{14}{24}} = \sqrt{\dfrac{7}{12}}$ Multiply radicands. Reduce.

(c) $3\sqrt{2} \cdot 4\sqrt{3} = 12\sqrt{6}$ Multiply coefficients 3 and 4. Then multiply radicands 2 and 3.

● **RELATED TOPIC: Multiplying fractions**

How to

Divide square-root radicals:

1. Divide coefficients to give the coefficient of the quotient.
2. Divide radicands to give the radicand of the quotient.
3. Simplify if possible.

Symbolically, $\dfrac{a\sqrt{b}}{c\sqrt{d}} = \dfrac{a}{c}\sqrt{\dfrac{b}{d}}$ c and $d \neq 0$

EXAMPLE   To divide the following radicals:

(a) $\dfrac{\sqrt{12}}{\sqrt{4}}$ (b) $\dfrac{\sqrt{\dfrac{2}{3}}}{\sqrt{\dfrac{7}{4}}}$ (c) $\dfrac{3\sqrt{6}}{6}$ (d) $\dfrac{5\sqrt{20}}{\sqrt{10}}$

(e) $\dfrac{8x^3\sqrt{9x^2}}{2x^2\sqrt{16x^3}}$

(a) $\dfrac{\sqrt{12}}{\sqrt{4}} = \sqrt{\dfrac{12}{4}} = \sqrt{3}$ Divide radicands.

(b) $\dfrac{\sqrt{\dfrac{2}{3}}}{\sqrt{\dfrac{7}{4}}} = \sqrt{\dfrac{\dfrac{2}{3}}{\dfrac{7}{4}}} = \sqrt{\dfrac{2}{3}\left(\dfrac{4}{7}\right)} =$ $\left(\dfrac{2}{3} \div \dfrac{7}{4} = \dfrac{2}{3} \cdot \dfrac{4}{7}\right)$

$\sqrt{\dfrac{8}{21}} = \sqrt{\dfrac{4 \cdot 2}{21}} = \dfrac{2\sqrt{2}}{\sqrt{21}}$ Factor the numerator to get a perfect-square factor; then, take the square root.

(c) $\dfrac{3\sqrt{6}}{6}$ The 3 and the denominator 6 can be divided (or reduced) because they are both outside the radical.

$\dfrac{\overset{1}{3}\sqrt{6}}{\underset{2}{6}} = \dfrac{\sqrt{6}}{2}$ A coefficient of 1 does not have to be written in front of the radical.

(d) $\dfrac{5\sqrt{20}}{\sqrt{10}}$ The 20 and 10 are divided because they are both square-root radicands.

$\dfrac{5\sqrt{20}}{\sqrt{10}} = 5\sqrt{2}$ Divide radicands.

(e) $\dfrac{8x^3\,\sqrt{9x^2}}{2x^2\,\sqrt{16x^3}}$ Reduce the factors outside the radical.

$\dfrac{4x\,\sqrt{9\,x^2}}{\sqrt{16\,x^3}}$ Reduce the factors inside the radical.

$\dfrac{4x\sqrt{9}}{\sqrt{16\,x}}$ Remove perfect-square factors from all radicands.

$\dfrac{4\,x(3)}{4\,\sqrt{x}}$ Reduce the resulting fraction.

$\dfrac{3x}{\sqrt{x}}$

● **RELATED TOPICS: Dividing fractions, Multiplying fractions**

The process of writing an expression with no radicals in the denominator is called **rationalizing the denominator.**

How to

Rationalize a denominator:

1. Reduce factors outside the radical.
2. Reduce factors inside the radical.
3. Remove perfect-square factors from all radicands.
4. Multiply the denominator by another radical so that the resulting radicand is a perfect square.
5. To preserve the value of the fraction, multiply the numerator by the same radical. Thus, we multiply by an equivalent of 1.
6. Reduce the resulting fraction, if possible.

EXAMPLE   To rationalize the denominator $\dfrac{1}{\sqrt{2}}$:

$\dfrac{1}{\sqrt{2}}\cdot\dfrac{\sqrt{2}}{\sqrt{2}} =$ Multiply by 1 in the form of $\dfrac{\sqrt{2}}{\sqrt{2}}$ so that the radicand in the denominator is a perfect square.

$\dfrac{\sqrt{2}}{\sqrt{4}} =$ Take the square root of the denominator.

$\dfrac{\sqrt{2}}{2}$

How to

Simplify a radical expression:

1. Reduce radicands and coefficients whenever possible.
2. Simplify expressions with perfect-square factors in all radicands.
3. Reduce radicands and coefficients whenever possible.
4. Rationalize denominators that contain radicals.
5. Reduce radicands and coefficients whenever possible.

EXAMPLE   To perform the operation and simplify if possible:

(a) $\sqrt{y^3} \cdot \sqrt{8y^2}$ (b) $\sqrt{\dfrac{1}{3}} \cdot \sqrt{\dfrac{8}{3x}}$

(a) $\sqrt{y^3} \cdot \sqrt{8y^2} = \sqrt{8y^5} = \sqrt{4 \cdot 2 \cdot y^4 \cdot y} = 2y^2 \sqrt{2y}$

(b) $\sqrt{\dfrac{1}{3}} \cdot \sqrt{\dfrac{8}{3x}} = \sqrt{\dfrac{8}{9x}} = \dfrac{\sqrt{8}}{\sqrt{9x}} = \dfrac{\sqrt{4 \cdot 2}}{\sqrt{9 \cdot x}} = \dfrac{2\sqrt{2}}{3\sqrt{x}} \cdot \dfrac{\sqrt{x}}{\sqrt{x}}$

$$= \dfrac{2\sqrt{2x}}{3x}$$

■ 14–4 Equations with Squares and Square Roots

A **quadratic equation** is an equation that has degree 2, that is, it has a squared variable.

How to

Solve an equation containing only squared variable terms and number terms:

1. Perform all normal steps to isolate the squared variable and to obtain a numerical coefficient of 1.
2. Take the square root of both sides.
3. Identify *both* roots of the equation and write them in set notation.

Roots and Radicals

EXAMPLE   To solve the equations:

    (a) $x^2 = 4$ (b) $x^2 + 9 = 25$ (c) $5x^2 + 10 = 30$

    (d) $\dfrac{3}{25} = \dfrac{x^2}{48}$

(a) $x^2 = 4$ Take the square root of both sides.

        $x = \pm 2$ Identify both roots.

        $x = +2$ or $x = -2$

        $\{2, -2\}$

(b) $x^2 + 9 = 25$ Isolate the squared variable term.

        $x^2 = 25 - 9$ Combine like terms.

        $x^2 = 16$ Take the square root of both sides.

        $x = \pm 4$ Identify both roots.

        $x = 4$ or $x = -4$

        $\{4, -4\}$

(c) $5x^2 + 10 = 30$ Isolate the squared variable term.

        $5x^2 = 30 - 10$ Combine like terms.

        $5x^2 = 20$ Divide by the coefficient of x^2.

        $\dfrac{5x^2}{5} = \dfrac{20}{5}$

        $x^2 = 4$ Take the square root of both sides.

        $x = \pm 2$ Identify both roots.

        $x = 2$ or $x = -2$

        $\{2, -2\}$

(d) $\dfrac{3}{25} = \dfrac{x^2}{48}$ To solve a proportion, cross multiply.

    $3(48) = 25x^2$

    $144 = 25x^2$ Divide by the coefficient of x^2.

    $\dfrac{144}{25} = \dfrac{25x^2}{25}$

$$\frac{144}{25} = x^2 \qquad \text{Take the square root of both sides.}$$

$$\pm\frac{12}{5} = x \qquad \text{Identify both roots.}$$

$$x = \frac{12}{5} \text{ or } x = -\frac{12}{5}$$

$$\left\{\frac{12}{5}, -\frac{12}{5}\right\}$$

⌐

● **RELATED TOPIC: Pure quadratic equations**

Just as taking the square root of both sides of an equation allows us to solve certain equations with squared variables, squaring both sides of an equation is used to solve certain types of equations with radical terms.

> **How to**
>
> *Solve an equation containing square-root radicals that are isolated on either or both sides of the equation:*
>
> **1.** Square both sides to eliminate any radicals.
> **2.** Perform all normal steps to isolate the variable and solve the equation.

EXAMPLE   To solve the following equations:

(a) $\sqrt{x} = \dfrac{3}{4}$ (b) $\sqrt{x - 2} = 5$ (c) $\sqrt{5x^2 - 36} = 8$

(d) $\sqrt{1.7x^2} = \sqrt{6.8}$

(a) $\sqrt{x} = \dfrac{3}{4}$ \qquad\qquad Square both sides of the equation.

$$\left(\sqrt{x}\right)^2 = \left(\frac{3}{4}\right)^2 \qquad \left(\sqrt{x}\right)^2 = x$$

$$x = \frac{9}{16} \text{ or } \left\{\frac{9}{16}\right\}$$

(b) $\sqrt{x - 2} = 5$ Square both sides of the equation.

$(\sqrt{x - 2})^2 = 5^2$ Any square-root radical squared equals the radicand.

$x - 2 = 25$ Isolate the variable.

$x = 25 + 2$ Combine like terms.

$x = 27$ or $\{27\}$

(c) $\sqrt{5x^2 - 36} = 8$

$(\sqrt{5x^2 - 36})^2 = 8^2$ Do this step mentally.

$5x^2 - 36 = 64$ Isolate variable term.

$5x^2 = 64 + 36$ Combine like terms.

$5x^2 = 100$ Divide by the coefficient of x^2.

$\dfrac{5x^2}{5} = \dfrac{100}{5}$ Do this step mentally.

$x^2 = 20$ Take square root of both sides.

$x = \pm\sqrt{20}$ Identify both roots.

$x = \pm2\sqrt{5}$ Exact roots.

$\{-2\sqrt{5}$ or $2\sqrt{5}\}$

$x \approx \pm4.472$ Approximate roots.

$\{-4.472$ or $4.472\}$

(d) $\sqrt{1.7x^2} = \sqrt{6.8}$ Square both sides to eliminate radicals.

$1.7x^2 = 6.8$ Divide by the coefficient of x^2.

$x^2 = \dfrac{6.8}{1.7}$ Simplify.

$x^2 = 4$ Take square root of both sides.

$x = \pm2$ or $\{+2, -2\}$ Identify both roots.

● **RELATED TOPICS: Linear equations, Quadratic equations**

When do I take the square root of both sides? When do I square both sides?

These are important questions.

- If an equation has a squared letter term, take the square root of both sides at the *end* of the solution. That is, when the squared letter has been isolated with a coefficient of +1, take the square root of both sides.
- If an equation has one or two radical terms but each radical is isolated on one side of the equal sign, square both sides of the equation at the *beginning* of the solution. Then, solve the remaining equation that contains no radicals.

Checking roots is always important

It is always advisable to check every root of an equation. It is especially important when you square both sides of an equation since this process can introduce an *extra* root. This extra root will not make a true statement when substituted in the original equation. Roots that do not make the equation true are called **extraneous roots**.

CHAPTER 15

Solving Quadratic and Higher-Degree Equations

- ■ **15–1 Solving Quadratic Equations**
- ■ **15–2 Solving Higher-Degree Equations**

■ 15–1 Solving Quadratic Equations

A **quadratic equation** is an equation in which at least one letter term is raised to the second power, and no letter term has a power higher than 2 or less than 0. The **standard form** for a quadratic equation is $ax^2 + bx + c = 0$, where a, b, and c are real numbers and $a > 0$.

The standard form of quadratic equations, $ax^2 + bx + c = 0$, has three types of terms.

1. ax^2 is a **quadratic term;** that is, the degree of the term is 2. In standard form, this term is the leading term and a is the leading coefficient.
2. bx is a **linear term;** that is, the degree of the term is 1 and the coefficient is b.
3. c is a **number term** or **constant term;** that is, the degree of the term is 0. The coefficient is c ($c = cx^0$).

A quadratic equation that has all three types of terms is sometimes referred to as a **complete quadratic equation.**

$ax^2 + bx + c = 0$ Complete quadratic equation.

A quadratic equation that has a quadratic term (ax^2) and a linear term (bx) but no constant term (c) is sometimes referred to as an **incomplete quadratic equation.**

$ax^2 + bx = 0$ Incomplete quadratic equation.

A quadratic equation that has a quadratic term (ax^2) and a constant term (c) but no linear term (bx) is sometimes referred to as a **pure quadratic equation,** the type that we worked with in Chapter 14.

$ax^2 + c = 0$ Pure quadratic equation.

The quadratic formula can be used to solve all types of quadratic equations.

Quadratic formula:

$$x = \frac{-b \pm \sqrt{b^2 - 4ac}}{2a}$$

where a, b, and c are coefficients of the terms of a quadratic equation in the form $ax^2 + bx + c = 0$.

How to

Solve a quadratic equation using the quadratic formula:

1. Set the equation equal to zero and arrange the equation in descending powers of the variable.
2. Identify the values for a, b, and c.
3. Substitute the values in the quadratic formula for a, b, and c.
4. Simplify the expression for the two solutions.

TIP

Common cause for errors

When you use the quadratic formula to solve problems, begin by writing the formula to help you remember it. When you write the formula, be sure to extend the fraction bar beneath the *entire* numerator. Not extending the fraction bar is a common cause for errors.

EXAMPLE   To solve $x^2 + 5x + 6 = 0$ for x using the quadratic formula:

$x^2 + 5x + 6 = 0$ The equation is already set equal to zero and is arranged in descending powers of x.

$a = 1$, $b = 5$, $c = 6$ Identify a, b, and c.

$$x = \frac{-b \pm \sqrt{b^2 - 4ac}}{2a}$$ Quadratic formula. Substitute for a, b, and c.

$$x = \frac{-5 \pm \sqrt{5^2 - 4 \cdot 1 \cdot 6}}{2 \cdot 1}$$ Square 5 and multiply in grouping.

$$x = \frac{-5 \pm \sqrt{25 - 24}}{2}$$ Combine terms under radical.

$$x = \frac{-5 \pm \sqrt{1}}{2}$$ Evaluate radical.

$$x = \frac{-5 \pm 1}{2}$$

At this point, we separate the formula into two parts, one using the $+1$ and the other using the -1.

$$x = \frac{-5 + 1}{2} \qquad\qquad x = \frac{-5 - 1}{2}$$

$$x = \frac{-4}{2} \qquad\qquad x = \frac{-6}{2}$$

$$x = -2 \qquad\qquad x = -3$$

Check $x = -2$ Check $x = -3$

$$x^2 + 5x + 6 = 0 \qquad x^2 + 5x + 6 = 0$$

$$(-2)^2 + 5(-2) + 6 = 0 \qquad (-3)^2 + 5(-3) + 6 = 0$$

$$4 - 10 + 6 = 0 \qquad 9 - 15 + 6 = 0$$

$$-6 + 6 = 0 \qquad -6 + 6 = 0$$

$$0 = 0 \qquad\qquad 0 = 0$$

The solution set is $\{-2, -3\}$.

How to

Solve a pure quadratic equation ($ax^2 + c = 0$):

1. Rearrange the equation, if necessary, so that quadratic or squared-letter terms are on one side and constants or number terms are on the other side of the equation.
2. Combine like terms, if appropriate.
3. Solve for the squared letter, so that the coefficient is +1.
4. Apply the square-root principle of equality by taking the square root of both sides.
5. Check both roots.

EXAMPLE   To solve $3y^2 = 27$:

$3y^2 = 27$ Divide both sides by the coefficient of the quadratic term.

$y^2 = 9$ Take the square root of both sides.

$y = \pm 3$

Check $y = +3$ Check $y = -3$

$3y^2 = 27$ $3y^2 = 27$

$3(\,3\,)^2 = 27$ $3(\,-3\,)^2 = 27$

$3(9) = 27$ $3(9) = 27$

$27 = 27$ $27 = 27$

The solution set is $\{3, -3\}$

A pure quadratic equation has two roots that are opposites.

● **RELATED TOPIC: Square-root principle of equality**

Zero-product property:

If the product of two variables is zero, one or both factors equal zero.
If $ab = 0$, then a or b or both equal zero.

How to

Solve an incomplete quadratic equation ($ax^2 + bx = 0$) by factoring:

1. If necessary, sort or transpose so that all terms are on one side of the equation in standard form, with zero on the other side.
2. Combine like terms.
3. Factor the common factor and set each factor equal to zero.
4. Solve for the variable in each equation formed in Step 3.

EXAMPLE To solve $2x^2 = 5x$ for x:

$$2x^2 = 5x \qquad \text{Write in standard form.}$$

$$2x^2 - 5x = 0 \qquad \text{Factor common factor } x.$$

$$x\,(2x - 5) = 0 \qquad \text{Write in factored form.}$$

$$x = 0 \quad 2x - 5 = 0 \qquad \text{Set each factor equal to zero.}$$

$$x = 0 \qquad 2x = 5 \qquad \text{Solve each equation.}$$

$$x = \frac{5}{2}$$

Check $x = 0$ Check $x = \dfrac{5}{2}$

$$2x^2 = 5x \qquad\qquad 2x^2 = 5x$$

$$2(\,0\,)^2 = 5(\,0\,) \qquad 2\left(\frac{5}{2}\right)^2 = 5\left(\frac{5}{2}\right)$$

$$2(0) = 5(0) \qquad \overset{1}{2}\left(\frac{25}{\underset{2}{4}}\right) = \frac{25}{2}$$

$$0 = 0 \qquad\qquad \frac{25}{2} = \frac{25}{2}$$

The solution set is $\left\{0, \frac{5}{2}\right\}$ or $\{0, 2.5\}$.

An incomplete quadratic equation has one root that is zero and one root that is not zero.

● **RELATED TOPIC: Common factors**

How to

Solve a complete quadratic equation ($ax^2 + bx + c = 0$)
by factoring:

1. If necessary, transpose, combine, and arrange terms so that all terms are on one side of the equation in standard form, with zero on the other side.
2. Factor the trinomial, if possible, into the product of two binomials.
3. Set each factor equal to zero.
4. Solve for the variable in each equation formed in Step 3.

EXAMPLE   To solve $6x^2 + 4 = 11x$ for x:

$$6x^2 + 4 = 11x \qquad \text{Write in standard form.}$$

$$6x^2 - 11x + 4 = 0 \qquad \text{The factors of 24 that add to } -11 \text{ are} \\ -3 \text{ and } -8. \text{ Write } -11x \text{ as } -3x - 8x.$$

$$6x^2 - 3x - 8x + 4 = 0 \qquad \text{Group.}$$

$$(6x^2 - 3x) + (-8x + 4) = 0 \qquad \text{Factor common factors in each grouping..}$$

$$3x\,(2x - 1) - 4\,(2x - 1) = 0 \qquad \text{Factor the common grouping.}$$

$$(3x - 4)\,(2x - 1) = 0 \qquad \text{Set each factor equal to 0 and solve for } x.$$

$$3x - 4 = 0 \qquad\qquad 2x - 1 = 0$$

$$3x = 4 \qquad\qquad 2x = 1$$

$$x = \frac{4}{3} \qquad\qquad x = \frac{1}{2}$$

Check $x = \dfrac{4}{3}$ $\qquad\qquad$ Check $x = \dfrac{1}{2}$

$$6x^2 + 4 = 11x \qquad\qquad 6x^2 + 4 = 11x$$

$$6\left(\frac{4}{3}\right)^2 + 4 = 11\left(\frac{4}{3}\right) \qquad\qquad 6\left(\frac{1}{2}\right)^2 + 4 = 11\left(\frac{1}{2}\right)$$

$$\overset{2}{6}\left(\frac{16}{\underset{3}{9}}\right) + 4 = \frac{44}{3} \qquad\qquad \overset{3}{6}\left(\frac{1}{\underset{2}{4}}\right) + 4 = \frac{11}{2}$$

$$\frac{32}{3} + \frac{12}{3} = \frac{44}{3} \qquad\qquad \frac{3}{2} + \frac{8}{2} = \frac{11}{2}$$

$$\frac{44}{3} = \frac{44}{3} \qquad\qquad \frac{11}{2} = \frac{11}{2}$$

L

● **RELATED TOPIC: Factoring trinomials**

Which method for solving quadratic equations is best?

A "best" method may be evaluated differently by different individuals. These suggestions are useful for solving quadratic equations.

1. Write the equation in standard form: $ax^2 + bx + c = 0$.
2. Identify the type of quadratic equation.
3. Solve pure quadratic equations by the square-root method.
4. Solve incomplete quadratic equations by finding common factors.
5. Solve complete quadratic equations by factoring into two binomials, if possible.
6. If factoring is not possible or is difficult, use the quadratic formula to solve the quadratic equation.

One way to determine the number of roots a quadratic equation has and the type of number the roots will be is to examine the **discriminant.** The discriminant is the radicand of the quadratic formula.

Properties of the discriminant, $b^2 - 4ac$:

1. If $b^2 - 4ac \geq 0$, the equation has real-number roots.
    a. If $b^2 - 4ac$ is a perfect square, there are two rational roots. The polynomial will factor.
    b. If $b^2 - 4ac = 0$, there is one rational root (sometimes called a **double root**).
    c. If $b^2 - 4ac$ is not a perfect square, there are two irrational roots.
2. If $b^2 - 4ac < 0$, the equation has no real-number roots. The roots are imaginary or complex.

EXAMPLE   To examine the discriminant of each equation and determine the number and type of the roots and to solve each equation:

(a) $5x^2 + 3x - 1 = 0$ (b) $3x^2 + 5x = 2$

(c) $4x^2 + 2x + 3 = 0$

(a) $5x^2 + 3x - 1 = 0$ $a = 5, b = 3, c = -1$

$\quad b^2 - 4ac = 3^2 - 4(5)(-1)$ Examine the
                                                                   discriminant.

$\qquad\qquad = 9 + 20$

$\qquad\qquad = \boxed{29}$ 29 is not a perfect
                                                                   square.

There will be two irrational roots.

$\quad x = \dfrac{-b \pm \sqrt{b^2 - 4ac}}{2a}$ Use the quadratic
                                                                        formula.

$\quad x = \dfrac{-3 \pm \sqrt{\boxed{29}}}{2(5)}$ Substitute 29 for the
                                                                        discriminant.

$\quad x = \dfrac{-3 + \sqrt{29}}{10}$ or $x = \dfrac{-3 - \sqrt{29}}{10}$ Exact irrational roots.

(b) $3x^2 + 5x = 2$ Write in standard form $a = 3, b = 5, c = -2$

$\quad 3x^2 + 5x - 2 = 0$

$\quad b^2 - 4ac = 5^2 - 4(3)(-2)$ Examine the discriminant.

$\qquad\qquad = 25 + 24$

$\qquad\qquad = \boxed{49}$ 49 is a perfect square.

There are two rational roots and the trinomial will factor.

$\quad (3x - 1)(x + 2) = 0$ Factor and set each factor equal to
                                                     zero.

$\quad 3x - 1 = 0 \qquad x + 2 = 0$ Solve each equation for x.

$\qquad 3x = 1$

$\qquad x = \dfrac{1}{3} \qquad x = -2$ Exact rational roots.

(c) $4x^2 + 2x + 3 = 0$ $a = 4, b = 2, c = 3$

$\quad b^2 - 4ac = 2^2 - 4(4)(3)$ Examine the discriminant.

$\qquad\qquad = 4 - 48$

$\qquad\qquad = \boxed{-44}$ $44 < 0$

There are no real roots. The roots are imaginary or complex.

$$x = \frac{-2 \pm \sqrt{-44}}{2 \cdot 4}$$ Quadratic formula. Simplify the radical.

$$x = \frac{-2 \pm 2i\sqrt{11}}{8}$$ Factor common factor.

$$x = \frac{2(-1 \pm i\sqrt{11})}{8}$$ Reduce.

$$x = \frac{-1 \pm i\sqrt{11}}{4}$$ The roots are complex.

● **RELATED TOPICS: Complex numbers, Imaginary numbers**

▲ **APPLICATION: Find the Dimensions of a Rectangle from Its Area**

How to

Solve an applied problem with a quadratic equation:

1. Identify the known values and let a variable equal the unknown.
2. Use the information to write a quadratic equation that represents the circumstances.
3. Select an appropriate method to solve the equation.

EXAMPLE To find the length and width of a rectangular table if the length is 8 inches more than the width and the area is 260 in². :

Unknown facts	Length and width of rectangle
Known facts	Area = 260 in², length = 8 in. more than width.
Relationships	Let x = number of inches in the width.
	$x + 8$ = number of inches in the length.
	Area = length times width, or $A = lw$.
Estimation	The square root of 260 is between 16 and 17; the width should be less than 16 and the length more than 16.

Calculations	$260 = (x + 8)(x)$	Substitute into area formula. Distribute.
	$260 = x^2 + 8x$	Write in standard form.
	$x^2 + 8x - 260 = 0$	
	$a = 1, \quad b = 8, \quad c = -260$	
	$x = \dfrac{-b \pm \sqrt{b^2 - 4ac}}{2a}$	Quadratic formula. Substitute.
	$x = \dfrac{-8 \pm \sqrt{(8)^2 - 4(1)(-260)}}{2(1)}$	Perform squaring and multiplying in radicand.
	$x = \dfrac{-8 \pm \sqrt{64 + 1{,}040}}{2}$	Combine terms in radicand.
	$x = \dfrac{-8 \pm \sqrt{1{,}104}}{2}$	Evaluate radicand.
	$x = \dfrac{-8 \pm 33.22649545}{2}$	Separate into two solutions.

$$x = \frac{-8 + 33.22649545}{2} \qquad x = \frac{-8 - 33.22649545}{2}$$

$$x = \frac{25.22649545}{2} \qquad\qquad x = \frac{-41.22649545}{2}$$

$$x = 12.6 \qquad\qquad\qquad x = -20.6$$

$$x + 8 = 20.6$$

Interpretation   Measurements are positive, so disregard the negative solution -20.6. **The width is 12.6 in. and the length is 20.6 in.**

⌐

● **RELATED TOPIC:** Area of a triangle

■ **15–2 Solving Higher-Degree Equations**

The **degree of an equation** in one variable is the highest power of any letter term that appears in the equation.

How to

Solve a higher-degree equation by factoring and using the zero product property:

1. Write the equation in standard form.
2. Factor the polynomial side of the equation.
3. Set each factor containing a variable equal to zero.
4. Solve each equation from Step 3.

EXAMPLE   To solve the equations by using the zero-product property:

(a) $x(x + 4)(x - 3) = 0$ (b) $2x^3 - 14x^2 + 20x = 0$

(c) $3x^3 - 15x = 0$

(a) $x\,(x + 4)\,(x - 3) = 0$ Already factored. Set
                                                                              each factor equal to zero.
         $x = 0$ $x + 4 = 0$ $x - 3 = 0$ Solve each equation.

         $x = 0$ $x = -4$ $x = 3$

The solution set is $\{0, -4, 3\}$.

(b) $2x^3 - 14x^2 + 20x = 0$ Factor common factors.

         $2x(x^2 - 7x + 10) = 0$ Factor the trinomial.

         $2x\,(x - 5)\,(x - 2) = 0$ Set each factor equal to
                                                                              zero.

         $2x = 0$ $x - 5 = 0$ $x - 2 = 0$ Solve each equation.

         $\dfrac{2x}{2} = \dfrac{0}{2}$ $x = 5$ $x = 2$

         $x = 0$

The solution set is $\{0, 5, 2\}$.

(c) $3x^3 - 15x = 0$ Factor common factors.

         $3x(x^2 - 5) = 0$ Set each variable factor equal to
                                                                      zero.

         $3x = 0$ $x^2 - 5 = 0$ Solve $x^2 - 5 = 0$ using the square
                                                                      root property.

         $\dfrac{3x}{3} = \dfrac{0}{3}$ $x^2 = 5$

$$x = 0 \qquad \sqrt{x^2} = \pm\sqrt{5}$$

$$x = \pm\sqrt{5}$$

The solution set is $\left\{0, \sqrt{5}, -\sqrt{5}\right\}$.

⬤ **RELATED TOPICS: Factoring, Square root property, Zero product property**

CHAPTER 16

Rational Expressions, Equations, and Inequalities

■ 16–1 Simplifying Rational Expressions

A **rational expression** is an algebraic fraction in which the numerator or denominator or both are polynomials.

> **How to**
>
> *Simplify rational expressions:*
>
> 1. Factor *completely* both the numerator and denominator.
> 2. Reduce factors common to both the numerator and denominator.
> 3. Write the simplified expression in either factored or expanded form.

Algebraic expressions that cannot be reduced are called **prime expressions**.

EXAMPLE   To reduce a rational expression to its simplest form:

(a) $\dfrac{x + y}{4x^2 + 4xy}$ (b) $\dfrac{a^2 + b^2}{a^2 - b^2}$ (c) $\dfrac{x^2 - 6x + 9}{x^2 - 9}$

(d) $\dfrac{3x^2 - 12}{6x + 12}$ (e) $\dfrac{a - b}{b - a}$

(a) $\dfrac{x + y}{4x^2 + 4xy} = \dfrac{(x + y)}{4x(x + y)} =$ Factor out the common factor in the denominator and consider the numerator as a grouping.

$\dfrac{(x + y)}{4x(x + y)} = \dfrac{1}{4x}$ Reduce. $x + y = 1(x + y)$

(b) $\dfrac{a^2 + b^2}{a^2 - b^2} = \dfrac{a^2 + b^2}{(a + b)(a - b)}$ Factor the difference of the squares in the denominator.

The numerator, which is the *sum* of the squares, will not factor. There are no factors common to both the numerator and denominator. Thus, the fraction is in simplest form:

$$\dfrac{a^2 + b^2}{(a + b)(a - b)} \quad \text{or} \quad \dfrac{a^2 + b^2}{a^2 - b^2}$$

(c) $\dfrac{x^2 - 6x + 9}{x^2 - 9} = \dfrac{(x - 3)(x - 3)}{(x + 3)(x - 3)}$ Write both the numerator and the denominator in factored form.

$= \dfrac{(x - 3)(x - 3)}{(x + 3)(x - 3)} = \dfrac{x - 3}{x + 3}$ Reduce common factors.

(d) $\dfrac{3x^2 - 12}{6x + 12} = \dfrac{3(x^2 - 4)}{6(x + 2)} = \dfrac{3(x + 2)(x - 2)}{(3)(2)(x + 2)}$ Factor the numerator and the denominator completely.

$= \dfrac{3(x + 2)(x - 2)}{(3)(2)(x + 2)} = \dfrac{x - 2}{2}$ Reduce common factor.

(e) $\dfrac{a - b}{b - a} =$ Write terms in numerator and denominator in same order.

$\dfrac{a - b}{-a + b} =$ Factor -1 in the denominator so that the leading coefficient of the binomial is positive.

$\dfrac{a - b}{-1(a - b)} =$ Reduce common factor.

$\dfrac{1}{-1} = -1$

TIP

Don't forget 1 or −1

In the preceding example, parts (a) and (e), some very important observations can be made.

- When all factors of a numerator or denominator reduce out, a factor of 1 remains.
- Polynomial factors are opposites when every term of one grouping is the opposite of a term of the other grouping.

    $(a - b)$ and $(b - a)$ are opposites.
    $(2m - 5)$ and $(5 - 2m)$ are opposites.

- Opposites differ by a factor of -1.

    $b - a = -1(a - b)$ $5 - 2m = -1(2m - 5)$

- When a numerator and denominator are opposites, the fraction reduces to -1.

$$\frac{a - b}{b - a} = \frac{a - b}{-1(a - b)} = \frac{1}{-1} = -1$$

$$\frac{2m - 5}{5 - 2m} = \frac{(2m - 5)}{-1(2m - 5)} = \frac{1}{-1} = -1$$

- It is helpful in recognizing common factors to rearrange terms in descending order and factor a -1 so that leading coefficients within a grouping are positive: $(5 - x) = (-x + 5)$ or $-1(x - 5)$.

Can common addends be reduced?

Does $\frac{5}{10}$ reduce to $\frac{3}{8}$? We know from previous experience that $\frac{5}{10}$ reduces to $\frac{1}{2}$ or 0.5. Then what is wrong with the following argument?

Both the numerator and the denominator of the fraction $\frac{5}{10}$ can be rewritten as addends or terms:

Does $\dfrac{5}{10} = \dfrac{2 + 3}{2 + 8} = \dfrac{3}{8}$?   **Can common addends be reduced? No!**

We can verify with our calculator that this is an incorrect statement, so the *process must be incorrect. Common addends cannot be reduced.* To correct the process, we reduce *only factors.* Thus, we rewrite $\frac{5}{10}$ as factors:

$$\frac{5}{10} = \frac{(1)(5)}{(2)(5)} = \frac{1}{2}$$

If common factors are reduced:

$$\frac{5}{10} = \frac{1}{2}$$

■ 16–2 Multiplying and Dividing Rational Expressions

How to

Multiply and divide rational expressions:

1. Convert any division to an equivalent multiplication.
2. Factor completely every numerator and denominator.
3. Reduce factors that are common to a numerator and denominator.
4. Multiply remaining factors.
5. The result can be written in factored or expanded form.

EXAMPLE   To perform the indicated operation and simplify:

(a) $\dfrac{x^2 - 4x - 12}{2x - 12} \cdot \dfrac{x - 4}{x^2 + 4x + 4}$ (b) $\dfrac{4y^2 - 9}{2y^2} \div \dfrac{y^2 - 2y - 15}{4y^2 + 12y}$

(a) $\dfrac{x^2 - 4x - 12}{2x - 12} \cdot \dfrac{x - 4}{x^2 + 4x + 4}$ Factor completely.

$= \dfrac{(x + 2)(x - 6)}{2(x - 6)} \cdot \dfrac{x - 4}{(x + 2)(x + 2)}$ Reduce common factors.

$= \dfrac{(x + 2)(x - 6)}{2(x - 6)} \cdot \dfrac{x - 4}{(x + 2)(x + 2)}$ Multiply remaining factors.

$= \dfrac{x - 4}{2(x + 2)}$ or $\dfrac{x - 4}{2x + 4}$ Factored or expanded form.

(b) $\dfrac{4y^2 - 9}{2y^2} \div \dfrac{y^2 - 2y - 15}{4y^2 + 12y}$ Convert division to multiplication.

$= \dfrac{4y^2 - 9}{2y^2} \cdot \dfrac{4y^2 + 12y}{y^2 - 2y - 15}$ Factor.

$= \dfrac{(2y + 3)(2y - 3)}{2y^2} \cdot \dfrac{4y(y + 3)}{(y + 3)(y - 5)}$ Reduce common factors.

$= \dfrac{(2y + 3)(2y - 3)}{2y^2} \cdot \dfrac{\overset{2}{4}y(y + 3)}{(y + 3)(y - 5)}$ Multiply.

$= \dfrac{2(2y + 3)(2y - 3)}{y(y - 5)}$ or $\dfrac{2(4y^2 - 9)}{y^2 - 5y}$ or $\dfrac{8y^2 - 18}{y^2 - 5y}$

A **complex rational expression** is a rational expression that has a rational expression in its numerator or its denominator or both. Examples of complex rational expressions are

$$\frac{\frac{4xy}{2x}}{5} \quad \text{and} \quad \frac{\frac{x^2 - y^2}{2x}}{\frac{x - y}{3x^2}}$$

● **RELATED TOPICS:** Complex fractions, Dividing fractions, Multiplying fractions

How to

Simplify a complex rational expression:

1. Rewrite the complex rational expression as a division of rational expressions.
2. Proceed to multiply or divide rational expressions.

TIP

Mentally convert from complex form to division and then to multiplication

Examine the symbolic representation for converting a complex expression to multiplication. The numerator is multiplied by the reciprocal of the denominator.

$$\frac{\frac{a}{b}}{\frac{c}{d}} = \frac{a}{b} \div \frac{c}{d} = \frac{a}{b} \cdot \frac{d}{c} \qquad b, c, \text{ and } d \neq 0.$$

When simplifying a complex rational expression, the rational expression can be written directly as a multiplication.

$$\frac{\frac{4xy}{2x}}{5} \quad \text{becomes} \quad \frac{4xy}{1} \cdot \frac{5}{2x} \qquad \begin{array}{l}\text{Numerator times reciprocal of}\\\text{denominator.}\end{array}$$

EXAMPLE   To simplify:

(a) $\dfrac{\dfrac{4xy}{2x}}{5}$ (b) $\dfrac{\dfrac{x^2 - y^2}{2x}}{\dfrac{x - y}{3x^2}}$

(a) $\dfrac{\dfrac{4xy}{2x}}{5} = \dfrac{4xy}{1} \cdot \dfrac{5}{2x}$

Multiply the numerator by the reciprocal of the denominator.

$= \dfrac{\overset{2}{4}xy}{1} \cdot \dfrac{5}{2x}$

Reduce then multiply.

$= \dfrac{10y}{1} = 10y$

Simplify.

(b) $\dfrac{\dfrac{x^2 - y^2}{2x}}{\dfrac{x - y}{3x^2}} = \dfrac{x^2 - y^2}{2x} \cdot \dfrac{3x^2}{x - y}$

Multiply the numerator by the reciprocal of the denominator.

$= \dfrac{(x + y)(x - y)}{2x} \cdot \dfrac{3x^2}{x - y}$

Factor.

$= \dfrac{(x + y)(x - y)}{2x} \cdot \dfrac{3\overset{x}{x^2}}{x - y}$

Reduce then multiply.

$= \dfrac{3x(x + y)}{2}$ or $\dfrac{3x^2 + 3xy}{2}$

Factored or expanded form.

⬤ **RELATED TOPIC: Reciprocal**

■ **16–3 Adding and Subtracting Rational Expressions**

How to

Add or subtract rational expressions with unlike denominators:

1. Find the *least common denominator (LCD)*.
2. Change *each* fraction to an equivalent fraction with the least common denominator.

3. Add or subtract numerators.
4. Keep the same denominator.
5. Reduce (or simplify) if possible.

EXAMPLE To subtract $\dfrac{x}{x-2} - \dfrac{5}{x+4}$:

Change each fraction to an equivalent fraction with a common denominator. The LCD is the product $(x-2)(x+4)$.

$$\frac{x}{x-2} = \frac{x(x+4)}{(x-2)(x+4)} = \frac{x^2+4x}{(x-2)(x+4)}$$

Multiply the numerator and denominator by $(x+4)$.

$$\frac{5}{x+4} = \frac{5(x-2)}{(x+4)(x-2)} = \frac{5x-10}{(x-2)(x+4)}$$

Multiply the numerator and denominator by $(x-2)$.

Use equivalent expressions and proceed.

$$\frac{x^2+4x}{(x-2)(x+4)} - \frac{5x-10}{(x-2)(x+4)} =$$

Subtract numerators.

$$\frac{x^2+4x-(5x-10)}{(x-2)(x+4)} =$$

Note: the *entire* numerator following the subtraction sign is subtracted.

$$\frac{x^2+4x-5x+10}{(x-2)(x+4)} =$$

Be careful with the signs when the parentheses are removed.

$$\frac{x^2-x+10}{(x-2)(x+4)} \quad \text{or} \quad \frac{x^2-x+10}{x^2+2x-8}$$

$x^2 - x + 10$ will not factor, so the fraction cannot be reduced.

⌞

● **RELATED TOPICS: Adding fractions, Least common denominator, Subtracting fractions**

Some complex rational expressions have terms in the numerator or denominator or both that should be combined to be a single term before proceeding to simplify.

EXAMPLE To simplify (a) $\dfrac{1+\dfrac{1}{x}}{\dfrac{2}{3}}$ and (b) $\dfrac{\dfrac{2}{x}-\dfrac{5}{2x}}{\dfrac{3}{4x}+\dfrac{3}{x}}$:

(a) $\dfrac{1 + \dfrac{1}{x}}{\dfrac{2}{3}} = $

Combine terms in the numerator.
LCD = x.

$\dfrac{\dfrac{x}{x} + \dfrac{1}{x}}{\dfrac{2}{3}} = $

Add like terms in the numerator.

$\dfrac{\dfrac{x + 1}{x}}{\dfrac{2}{3}} = $

Multiply the numerator by the reciprocal of the denominator.

$\dfrac{x + 1}{x} \cdot \dfrac{3}{2} = $

Multiply.

$\dfrac{3(x + 1)}{2x}$ or $\dfrac{3x + 3}{2x}$

(b) $\dfrac{\dfrac{2}{x} - \dfrac{5}{2x}}{\dfrac{3}{4x} + \dfrac{3}{x}} = $

Find the common denominator in the numerator (LCD = $2x$) and in the denominator (LCD = $4x$).

$\dfrac{\dfrac{4}{2x} - \dfrac{5}{2x}}{\dfrac{3}{4x} + \dfrac{12}{4x}} = $

$\dfrac{2}{x} \cdot \dfrac{2}{2} = \dfrac{4}{2x}, \quad \dfrac{3}{x} \cdot \dfrac{4}{4} = \dfrac{12}{4x}$
Subtract like terms in the numerator and add like terms in the denominator.

$\dfrac{\dfrac{-1}{2x}}{\dfrac{15}{4x}} = $

Multiply the numerator by the reciprocal of the denominator.

$\dfrac{-1}{2\overset{}{\underset{1}{x}}} \cdot \dfrac{\overset{2}{4x}}{15}$

Reduce.

$\dfrac{-2}{15}$ or $-\dfrac{2}{15}$

Multiply.

■ 16–4 Solving Equations with Rational Expressions

Excluded values of rational equations are values that would cause the expressions to be undefined.

Expressions
Inequalities

How to

Find excluded values of rational equations:

1. Set each denominator containing a variable equal to zero and solve for the variable.
2. Each equation in Step 1 produces an excluded value; however, some values may be repeats.

EXAMPLE   To determine the value or values that must be excluded as possible solutions:

(a) $\dfrac{1}{x} + \dfrac{1}{2} = 5$ (b) $\dfrac{1}{x} = \dfrac{1}{x-3}$

(a) $\dfrac{1}{x} + \dfrac{1}{2} = 5$ Set each denominator containing a variable equal to 0.

$x = 0$ Excluded value.

(b) $\dfrac{1}{x} = \dfrac{1}{x-3}$ Set each denominator containing a variable equal to 0.

$x = 0$, $x - 3 = 0$ Solve each equation.

$x = 3$

Excluded values are 0 and 3.

TIP

Excluded values

Because division by zero is undefined, any value of the variable that makes the value of any denomination zero is excluded as a possible solution or root of an equation. Therefore, it is important (1) to determine which values are *excluded values* and cannot be used as possible solutions and (2) to check each possible solution.

How to

Solve rational equations:

1. Determine the excluded values.

2. Clear the equation of all denominators by multiplying the entire equation by the *least common multiple* (LCM) of the denominators.
3. Complete the solution of the equation.
4. Eliminate any excluded values as solutions.
5. Check the solutions.

● **RELATED TOPICS:** Least common multiple, Solving linear equations

EXAMPLE   To solve the rational equations and check:

(a) $\dfrac{2}{y} + \dfrac{1}{3} = 1$ (b) $\dfrac{5}{x-2} - \dfrac{3x}{x-2} = -\dfrac{1}{x-2}$

(a) $\dfrac{2}{y} + \dfrac{1}{3} = 1$ Excluded value: $y = 0$.

LCM for y and 3 is $3y$.

$$3y\left(\dfrac{2}{y}\right) + 3y\left(\dfrac{1}{3}\right) = 3y\,(1)$$ Multiply by LCM, $3y$.

$$(3y)\left(\dfrac{2}{y}\right) + (3y)\left(\dfrac{1}{3}\right) = 3y(1)$$ Reduce and multiply remaining factors.

$$6 + y = 3y$$ Sort terms.

$$6 = 3y - y$$ Combine like terms.

$$6 = 2y$$ Divide by the coefficient of x.

$$\dfrac{6}{2} = \dfrac{2y}{2}$$

$$3 = y$$ Check because 3 is not an excluded value.

Check: $\dfrac{2}{y} + \dfrac{1}{3} = 1$

$$\dfrac{2}{3} + \dfrac{1}{3} = 1$$ Substitute 3 for y.

$$\dfrac{3}{3} = 1$$

$$1 = 1$$ Solution checks.

(b) $\dfrac{5}{x-2} - \dfrac{3x}{x-2} = -\dfrac{1}{x-2}$

Excluded value: $x - 2 = 0$
$x = 2$

$$(x - 2)\frac{5}{x - 2} - (x - 2)\frac{3x}{x - 2} = (x - 2)\left(-\frac{1}{(x - 2)}\right)$$

Multiply by LCM, $x - 2$, to clear the common denominator.

$$(x - 2)\frac{5}{x - 2} - (x - 2)\frac{3x}{x - 2} = (x - 2)\left(-\frac{1}{(x - 2)}\right)$$

Reduce.

$$5 - 3x = -1 \qquad \text{Sort terms.}$$

$$-3x = -1 - 5 \qquad \text{Combine like terms.}$$

$$-3x = -6 \qquad \text{Divide by the coefficient of } x.$$

$$\frac{-3x}{-3} = \frac{-6}{-3}$$

$$x = 2$$

This is an excluded value. $x = 2$ causes the denominator to be zero.

The equation has no solution.

Check: $$\dfrac{5}{x - 2} - \dfrac{3x}{x - 2} = -\dfrac{1}{x - 2} \qquad \text{Substitute 2 for } x.$$

$$\frac{5}{2 - 2} - \frac{3(2)}{2 - 2} = -\frac{1}{2 - 2}$$

$$\frac{5}{0} - \frac{6}{0} = -\frac{1}{0} \qquad \begin{array}{l}\text{Division by zero is}\\ \text{undefined. The equation}\\ \text{has no solution.}\end{array}$$

⌞

▲ **APPLICATION: Calculating Unit Costs**

EXAMPLE   At the January office products sale, Erma Thornton Braddy purchased a package of 3.5-in. computer disks for $16 and a package of writing pens for $6. He paid 40 cents more per pen than he paid for each computer disk, and the computer disk package contained 15 more disks than the pen package. To find the number of disks and pens he purchased and the price of each disk and pen:

Unknown facts   Let n = the number of pens. The number of pens, the number of disks, the cost per pen, and the cost per disk are all unknown.

Known facts   Disks cost $16 per package. Pens cost $6 per package. There are 15 more disks than pens. Each pen costs 40¢ more than each disk.

Relationships

Number of disks $= n + 15$

Cost per item $=$ Cost $\div$ Number of items

Cost per pen $= \dfrac{\$6}{n}$

Cost per disk $= \dfrac{\$16}{n + 15}$

Cost per pen $=$ Cost per disk $+ 40\cancel{c}$.

$40\cancel{c} = \dfrac{40}{100}$ dollars or \$0.40.

$\dfrac{6}{n} = \dfrac{16}{n + 15} + \dfrac{40}{100}$

Estimation

Pens cost more than disks and only \$6 was spent for pens, so there will be only a small number of pens.

$\dfrac{6}{n} = \dfrac{16}{n + 15} + \dfrac{40}{100} \qquad \dfrac{40}{100} = \dfrac{2}{5}$

$\dfrac{6}{n} = \dfrac{16}{n + 15} + \dfrac{2}{5}$

Calculations

Excluded values: $n = 0$ and $n + 15 = 0$ or $n = -15$. These excluded values would also be eliminated because they are inappropriate within the context of the problem.

$(n)(n + 15)(5)\,\dfrac{6}{n} = (n)(n + 15)(5)\,\dfrac{16}{n + 15} + (n)(n + 15)(5)\,\dfrac{2}{5}$

Multiply by LCM, $(n)(n + 15)(5)$.

$(\cancel{n})(n + 15)(5)\,\dfrac{6}{\cancel{n}} = (n)(\cancel{n + 15})(5)\,\dfrac{16}{\cancel{n + 15}} + (n)(n + 15)(\cancel{5})\,\dfrac{2}{\cancel{5}}$

Reduce.

$(n + 15)(5)(6) = (n)(5)(16) + (n)(n + 15)(2)$

Simplify coefficients.

$30(n + 15) = 80n + 2n(n + 15)$

Distribute.

$30n + 450 = 80n + 2n^2 + 30n$

Move all terms to one side.

$0 = 80n + 2n^2 + 30n - 30n - 450$

Combine like terms and write in descending order.

$$0 = 2n^2 + 80n - 450$$

Write in factored form by factoring a common factor.

$$0 = 2(n^2 + 40n - 225)$$

Factor trinomial.

$$0 = 2(n - 5)(n + 45)$$

Set variable factors equal to 0.

$$n - 5 = 0 \qquad n + 45 = 0$$

Solve.

$$n = 5 \text{ pens} \qquad n = -45$$

Disregard the negative root.

Interpretation   There are 5 pens, and that fact can be used to find the other missing facts.

$$n + 15 = 5 + 15 = 20 \text{ disks}$$

$$\frac{6}{n} = \frac{\$6}{5} = \$1.20 \text{ per pen}$$

$$\frac{16}{n + 15} = \frac{\$16}{20} = \$0.80 \text{ per disk}$$

Check: $\dfrac{6}{n} = \dfrac{16}{n + 15} + \dfrac{2}{5}$ Substitute 5 for n.

$$\frac{6}{5} = \frac{16}{5 + 15} + \frac{2}{5}$$

$$\frac{6}{5} = \frac{16}{20} + \frac{2}{5}$$ Change $\frac{2}{5}$ to an equivalent fraction.

$$\frac{6}{5} = \frac{16}{20} + \frac{8}{20}$$ Add fractions on the right side of the equal sign.

$$\frac{6}{5} = \frac{24}{20} \qquad\qquad \frac{24}{20} = \frac{6}{5}$$

$$\frac{6}{5} = \frac{6}{5}$$ The solution checks.

■ 16–5 Solving Quadratic and Rational Inequalities

The **critical values** or **boundaries** of a quadratic inequality are the values that make the quadratic *equation* associated with the quadratic inequality true.

How to

Solve quadratic inequalities by factoring:

1. Rearrange the inequality in standard form so the right side of the inequality is zero.
2. Write the left side of the inequality in factored form.
3. Determine the critical values by finding the values that make each factor equal to zero.
4. Plot the critical values on a number line to form the three regions.
5. Test each region of values by selecting any point within a region, substituting that value into the original inequality, solving the inequality, and deciding if the resulting inequality is a *true* statement.
6. The solution set for the quadratic inequality is the region or regions that produce a *true* statement in Step 5.
7. The critical values or boundary points are included in the solution set if the inequality is "inclusive" ($\leq$ or $\geq$). The critical values or boundary points are not included in the solution set if the inequality is "exclusive" ($<$ or $>$).

EXAMPLE To solve the inequality $x^2 + 5x + 6 \leq 0$:

$(x + 3)(x + 2) \leq 0$ Write in factored form.

$x + 3 = 0$ $x + 2 = 0$ Determine the critical values.

$x = \boxed{-3}$ $x = \boxed{-2}$ Plot the critical values and label the corresponding regions.

Region I   Region II   Region III

-4 -3 -2 -1 0

Region I: $x \leq -3$ To the left of the smaller critical value, -3.

Region II: $-3 \leq x \leq -2$ Between the critical values, -3 and -2.

Region III: $x \geq -2$ To the right of the larger critical value, -2.

Test each region.

Region I: $x \leq -3$	Region II: $-3 \leq x \leq -2$
Region I test point: $x = -4$	**Region II test point:** $x = -2.5$
$(x + 3)(x + 2) \leq 0$	$(-2.5 + 3)(-2.5 + 2) \leq 0$
$(-4 + 3)(-4 + 2) \leq 0$	$(0.5)(-0.5) \leq 0$
$(-1)(-2) \leq 0$	$-0.25 \leq 0$
$2 \leq 0$	

The inequality is false, so Region I is not in the solution set.

The inequality is true, so Region II is in the solution set.

Region III: $x \geq -2$

Region III test point: $x = -1$

$$(-1 + 3)(-1 + 2) \leq 0$$

$$(2)(1) \leq 0$$

$$2 \leq 0$$

The inequality is false, so Region III is not in the solution set.

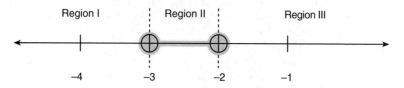

The solution set is $-3 \leq x \leq -2$ or $[-3, -2]$.

● **RELATED TOPIC: Graphing linear inequalities in one variable**

Selecting test points

Boundary points should not be used as test points. When possible, select integers as test points, and if zero is in a region, it makes an excellent test point.

Quadratic inequalities can also be solved by analyzing the regions formed by the inequalities of the factors. The previous example is solved below.

$x^2 + 5x + 6 \leq 0$ Original inequality.

$(x + 3)(x + 2) \leq 0$ Factored form of inequality.

Critical values: $x = -3$ and $x = -2$

The critical value of a factor is the point on the number line where the factor equals zero. Negative values of the factor are to the left of the critical value and positive values are to the right.

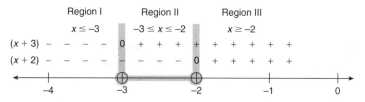

The product $(x + 3)(x + 2)$ is *positive* where the signs of the factors are *alike*, Regions I and III.

The product $(x + 3)(x + 2)$ is *negative* where the signs of the factors are *unlike*, Region II.

The factored form of the given inequality is "less than or equal to" zero. Therefore, the **solution set includes the region or regions where the product is *negative* since the inequality is less than zero.**

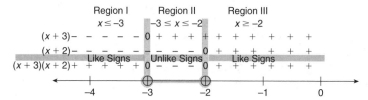

The solution set is $-3 \leq x \leq -2$ or $[-3, -2]$.

Finding solution sets by inspection for quadratic inequalities

Once you understand how to identify the appropriate region or regions for the solution set of a quadratic inequality, you may observe a pattern that allows you to select appropriate regions for such inequalities by inspection. Patterns and generalizations are appropriate only under special conditions. Here, the inequality **must** be written in standard form ($ax^2 + bx + c < 0$ or $ax^2 + bx + c > 0$) where $a > 0$ (a is positive). The critical values are represented symbolically as s_1 and s_2 where $s_1 < s_2$. The critical values and signs of the two factors and the product are shown below.

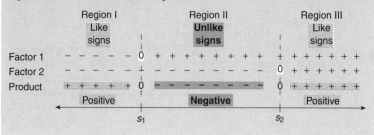

When the inequality $ax^2 + bx + c < 0$ is written in factored form, the two factors must have opposite signs to make a product that is less   than zero (negative). The solution set is in Region II and is written symbolically as $s_1 < x < s_2$. The graphical representation of this solution appears below on the left.

When the inequality $ax^2 + bx + c > 0$ is written in factored form, the two factors must have the same signs to make a product that is greater   than zero (positive). We see that the solution set is in Regions I and III and is written symbolically as $x < s_1$ or $x > s_2$. The graphical representation of this solution appears in the figure on the right.

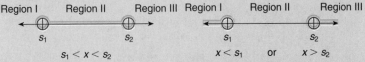

Region I   Region II   Region III Region I   Region II   Region III

$$s_1 < x < s_2$$ $$x < s_1 \quad \text{or} \quad x > s_2$$

Now, if you solve a quadratic equation by *any* method, including using the quadratic formula, you can determine the solution set of the associated quadratic inequality by inspection after you have found the critical values. The solution can be checked by testing one point from the solution set. Similar generalizations can be made for $ax^2 + bx + c \leq 0$ and $ax^2 + bx + c \geq 0$ for $a > 0$.

● **RELATED TOPICS:** Absolute-value inequalities, Linear inequalities

A **rational inequality** is an inequality that contains at least one rational (fraction) term.

How to

Solve a rational inequality like $\frac{x + a}{x + b} < 0$ or $\frac{x + a}{x + b} > 0$:

1. Find critical values by setting the numerator and denominator equal to zero.
2. Solve each equation from Step 1 and use the solutions (critical values) to divide the number line into three regions.
3. Regions that have like signs for both the numerator and denominator are in the solution set of $\frac{x + a}{x + b} > 0$ for $x \neq -b$ ($x = -b$ is an excluded value).
4. Regions that have unlike signs for the numerator and denominator are in the solution set of $\frac{x + a}{x + b} < 0$ for $x \neq -b$ ($x = -b$ is an excluded value).

EXAMPLE   To solve the following inequality:

$$\frac{x + 2}{x - 5} < 0$$ Set numerator and denominator equal to zero.

$x + 2 = 0$ $x - 5 = 0$ Solve each equation.

$x = -2$ $x = 5$ Critical values for the inequality.

The quotient in Region I is positive (like signs). The quotient in Region II is negative (unlike signs). The quotient in Region III is positive (like signs).

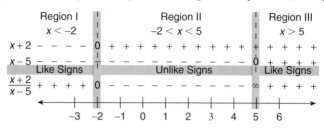

Since the given inequality is less than zero (negative) the solution is

$-2 < x < 5$ or $(-2, 5)$ or

■ 16–6 Graphing Quadratic Equations and Inequalities in Two Variables

The graphs of linear equations are *straight* lines. The graph of an equation of a degree higher than 1 is a *curved* line. The graph of a quadratic equation in two variables in the form $y = ax^2 + bx + c$ is a **parabola.**

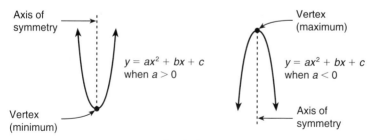

Parabolas

A parabola is symmetrical; that is, it can be folded in half and the two halves match. The fold line is called the **axis of symmetry.** For a parabola in the form $y = ax^2 + bx + c$, the equation of the axis of symmetry is $x = -\frac{b}{2a}$.

The point of the graph that crosses the axis of symmetry is the **vertex** of the parabola. Thus, the x-coordinate of the vertex of the parabola is $-\frac{b}{2a}$.

The vertex of the parabola is the point where the parabola reaches its **maximum** or **minimum** value.

How to

Graph quadratic equations in the form $y = ax^2 + bx + c$:

1. Find the axis of symmetry: $x = -\frac{b}{2a}$.
2. Find the vertex: x-coordinate of vertex $= -\frac{b}{2a}$. To find the y-coordinate of the vertex, substitute the x-coordinate into the original equation and solve for y.
3. Find one or two additional points that are to the right of the axis of symmetry.
4. Apply the property of symmetry to find additional points to the left of the axis of symmetry.
5. Connect the plotted points with a smooth, continuous curved line.

EXAMPLE To graph $y = x^2 - 6x + 8$:

To find the axis of symmetry:

$$x = -\frac{b}{2a} \qquad \text{Substitute values.}$$

$$x = -\frac{-6}{2(1)} \qquad \text{Evaluate.}$$

$$x = \frac{6}{2}$$

$$x = 3$$

Axis of symmetry: $x = 3$

To find the vertex:

$$y = (3)^2 - 6(3) + 8 \qquad \text{Substitute } x = 3.$$

$$y = 9 - 18 + 8$$

$$y = -1$$

Vertex: $(3, -1)$

To find the x-intercepts:

$$0 = x^2 - 6x + 8 \qquad \text{Substitute } y = 0.$$

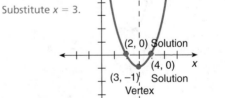

$$0 = (x - 2)(x - 4)$$

$$x - 2 = 0 \qquad x - 4 = 0$$

$$x = 2 \qquad x = 4$$

x-intercepts: $(2, 0)$ and $(4, 0)$

The graph of a quadratic inequality in two variables has a boundary that is the graph of a similar equation. Again, the solid line or dashed line is used to show if the boundary is included or excluded. The solution set of a quadratic inequality in the form of a parabola is either all points inside the parabola or all points outside the parabola.

How to

Graph a quadratic inequality in two variables:

1. Find the boundary by graphing an equation that substitutes an equal sign for the inequality symbol (boundary included-solid line; boundary excluded, dashed line).
2. Test any point that is not on the boundary line.
3. If the test point makes a true statement with the original inequality, shade the portion of the parabola containing the test point.
4. If the test point makes a false statement with the original inequality, shade the opposite portion of the parabola.

EXAMPLE   To graph $y \leq x^2 - 6x + 8$:
Graph the boundary $y = x^2 - 6x + 8$
(See previous example.)

Test point: $(0, 0)$

$y \leq x^2 - 6x + 8$
Substitute.

$0 \leq 0^2 - 6(0) + 8$

$0 \leq 8$
True.

Shade outside the parabola.

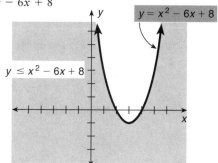

CHAPTER 17

Exponential and Logarithmic Equations

- ■ **17–1 Exponential Expressions and Equations**
- ■ **17–2 Logarithmic Expressions and Equations**

■ 17–1 Exponential Expressions and Equations

Many formulas have terms that contain exponents. An **exponential expression** is an expression that contains at least one term that has a **variable exponent**. A variable exponent is an exponent that has at least one letter factor. Exponential expressions can be evaluated on a calculator or computer by using the general power key $\boxed{x^y}$ or $\boxed{\wedge}$ or other power keys as appropriate. An **exponential equation** or formula contains at least one term that has a variable exponent.

A commonly used formula that contains a variable exponent is the formula for calculating the compound amount for compound interest. **Compound interest** is the interest calculated at the end of each period and then added to the principal for the next period. The **compound amount** or the **accumulated amount** is the principal and interest accumulated over a period of time.

Compound amount (Accumulated amount):

$$A = P\left(1 + \frac{r}{n}\right)^{nt}$$

where A = accumulated amount $\qquad P$ = original principal

t = time in years $\qquad r$ = rate per year

n = number of compounding periods per year

EXAMPLE   Use the formula $A = P(1 + \frac{r}{n})^{nt}$ and a calculator or a computer, to find the accumulated amount on an investment of $1,500, invested at an interest rate of 9% for 3 years, if the interest is compounded quarterly.

Estimation   We expect to have more than $1,500.

$$A = P\left(1 + \frac{r}{n}\right)^{nt}$$ $P = \$1,500;\ r = 9\%$ or $0.09;\ n =$ quarterly or 4 times a year; $t = 3$ years.

$$A = 1,500\left(1 + \frac{0.09}{4}\right)^{(4)(3)}$$ Simplify exponent and division term in grouping.

$$A = 1,500(1 + 0.0225)^{12}$$ Combine terms in grouping.

$$A = 1,500(1.0225)^{12}$$ $1.0225\ \boxed{\wedge}\ 12\ \boxed{=}\ \Rightarrow$ 1.30604999

$$A = 1,500(1.30604999)$$ Multiply.

$$A = 1,959.07$$ Rounded.

Interpretation   **The accumulated amount of the $1,500 investment after 3 years is $1,959.07 to the nearest cent.**

⌊

● **RELATED TOPIC: Order of operations**

▲ **APPLICATION: Maximize Your Investments: Start Early!**

Many young adults do not realize the importance of starting an investment strategy early. The greater benefits of compound interest are realized over longer periods of time. For instance, suppose by age 35 you have accumulated **$25,000** in savings and invest it in a long-term program that is compounded annually at 8%. At age 65, this investment will have accumulated to $251,566.

EXAMPLE   Investment value of $25,000 after 30 years:

$$A = P\left(1 + \frac{r}{n}\right)^{nt}$$

$P = \$25,000;\ r = 0.08;\ n = 1;\ t = 30$ Substitute values.

$$A = \$25,000\left(1 + \frac{0.08}{1}\right)^{1(30)}$$ Simplify grouping.

$$A = \$25,000(1.08)^{30}$$ Raise to the power, then multiply.

$$A = \$251,566.42$$

If it takes you until age 45 to accumulate **$25,000** in savings and you invest it in a long-term program that is compounded annually at 8%, at age 65 this investment will have accumulated to only $116,524.

Investment value of $25,000 after 20 years:

$$A = P\left(1 + \frac{r}{n}\right)^{nt}$$

$P = \$25,000; r = 0.08; n = 1; t = 20$ Substitute values.

$$A = 25,000\left(1 + \frac{0.08}{1}\right)^{1(20)}$$ Simplify grouping and exponent.

$A = 25,000(1.08)^{20}$ Raise to the power then multiply.

$A = \$116,523.93$ Rounded.

Even if you have accumulated **$50,000** by age 45 and invest it in a long-term program that is compounded annually at 8%, this investment will have accumulated to only $233,048 by the time you are 65.

Investment value of $50,000 after 20 years:

$$A = P\left(1 + \frac{r}{n}\right)^{nt}$$

$P = \$50,000; r = 0.08; n = 1; t = 20$ Substitute values.

$$A = 50,000\left(1 + \frac{0.08}{1}\right)^{1(20)}$$

$A = 50,000(1.08)^{20}$

$A = \$233,047.86$ Rounded.

Even a modest investment plan, started early and continued consistently, will reap significant benefits.

The accumulated amount is also the **future value** or **maturity value.** Many business calculator or computer software programs have a future value function *(FV)*.

● **RELATED TOPICS: Future value of an annuity due, Future value of an ordinary annuity, Order of operations**

TIP

Interpreting a formula

In working with formulas it is important to understand what is represented by each letter of the formula. The compound interest formula can also be given in terms of the number of **compounding periods** and the **interest rate per period.**

$$\text{total compounding periods } (N) = \begin{array}{c}\text{compounding periods per year } (n) \\ \text{times} \\ \text{number of years } (t)\end{array}$$

$$N = nt$$

$$\text{interest rate per period } (R) = \frac{\text{interest rate per year } (r)}{\text{compounding periods per year } (n)}$$

$$R = \frac{r}{n}$$

The compound amount or future value formula can also be written as

$$A = P(1 + R)^N \quad \text{or} \quad FV = P(1 + R)^N$$

where A or FV = accumulated amount or future value

R = interest rate per compounding period

N = total number of compounding periods

▲ **APPLICATION: Compound Interest on an Investment**

How to

Find the compound interest:

1. Find the accumulated amount or future value using the formula

$$A = P(1 + R)^N$$

2. Subtract the original principal from the accumulated amount.

$$I = A - P$$

Combining both formulas:

$$I = P(1 + R)^N - P$$

or

$$I = P\big((1 + R)^N - 1\big)$$

EXAMPLE   To find the compound interest on an investment of $10,000 at 8% annual interest compounded semiannually for 3 years:

$$P = 10,000, R = \frac{0.08}{2} = 0.04, N = 2 \times 3 = 6 \text{ periods}$$

$I = P(1 + R)^N - P$	Substitute values.
$I = 10,000(1 + 0.04)^6 - 10,000$	Combine terms inside grouping.
$I = 10,000(1.04)^6 - 10,000$	Raise to the power.
$I = 10,000(1.265319018) - 10,000$	Multiply.
$I = 12,653.19 - 10,000$	Subtract.
$I = \$2,653.19$	

The interest is $2,653.19.

● **RELATED TOPICS: Compound amount, Simple interest**

The **present value** of an investment is the **lump sum** amount that should be invested now at a given interest rate for a specific period of time to yield a specific accumulated amount in the future.

Present value

$$PV = \frac{FV}{(1 + R)^N}$$

where PV = present value

FV = future value

R = interest rate per period

N = total number of periods

▲ **APPLICATION: Present Value of a Lump Sum Investment**

EXAMPLE   The 7th Inning Sports Shop needs $20,000 in 10 years to replace engraving equipment. To find the amount the firm must invest at the present if it receives 10% interest compounded annually:

$$FV = 20,000, R = \frac{0.10}{1} = 0.1, N = 10(1) = 10$$

$$PV = \frac{20,000}{(1 + 0.1)^{10}}$$ Combine terms inside grouping.

$$PV = \frac{20,000}{(1.1)^{10}}$$ Raise denominator to the power.

$$PV = \frac{20,000}{2.59374246}$$ Divide.

$$PV = \$7,710.87$$

$7,710.87 should be invested now.

● **RELATED TOPIC: Present value of an ordinary annuity**

In advertising or in stating the term of an investment or loan it is common to equate the compound interest rate to a comparable simple interest rate. This rate is referred to as the **effective rate, annual percentage rate (APR)** and **annual percentage yield (APY).**

How to

Find the effective rate:

$$E = \left(1 + \frac{r}{n}\right)^n - 1$$

where E = effective rate

r = interest rate per year

n = number of compounding periods per year

▲ **APPLICATION: Effective Interest Rate for a Loan**

EXAMPLE   To find the effective interest rate for a loan of $600 at 10% compounded semiannually:

$r = 0.10; n = 2$

$$E = \left(1 + \frac{r}{n}\right)^n - 1$$ Substitute values.

$$E = \left(1 + \frac{0.1}{2}\right)^2 - 1 \qquad \text{Simplify inside grouping.}$$

$$E = (1 + 0.05)^2 - 1$$

$$E = (1.05)^2 - 1 \qquad \text{Raise to the power.}$$

$$E = 1.1025 - 1 \qquad \text{Subtract.}$$

$$E = 0.1025$$

Effective interest rate is 10.25%.

The Natural Exponential, e

Exponential change is an interesting phenomenon. Let's look at the value of the expression $\left(1 + \frac{1}{n}\right)^n$ as n gets larger and larger.

Values of $\left(1 + \frac{1}{n}\right)^n$		
n	$\left(1 + \frac{1}{n}\right)^n$	Result
1	$\left(1 + \frac{1}{1}\right)^1$	2
2	$\left(1 + \frac{1}{2}\right)^2$	2.25
3	$\left(1 + \frac{1}{3}\right)^3$	2.37037037
100	$\left(1 + \frac{1}{100}\right)^{100}$	2.704813829
1,000	$\left(1 + \frac{1}{1,000}\right)^{1,000}$	2.716923932
10,000	$\left(1 + \frac{1}{10,000}\right)^{10,000}$	2.718145927
100,000	$\left(1 + \frac{1}{100,000}\right)^{100,000}$	2.718268237
1,000,000	$\left(1 + \frac{1}{1,000,000}\right)^{1,000,000}$	2.718280469

The value of the expression changes very little as the value of n gets larger. We can say that the value approaches a given number. We call the given number *e,* the **natural exponential.** The natural exponential, *e,* like π, is an irrational number and will never terminate or repeat as more decimal places are examined.

The natural exponential, *e,* like π, is a *constant.* That is, the value is always the same; it does not vary. The **natural exponential, *e,*** is the limit that the value of the expression $\left(1 + \frac{1}{n}\right)^n$ approaches as n gets larger and larger without bound. The value of *e* to nine decimal places is 2.718281828.

To evaluate formulas containing the natural exponential, *e,* we can use a calculator or computer. The function key is most often labeled $\boxed{e^x}$.

▲ APPLICATION: Compound Amount of Continuous Compound Interest

As the number of compounding periods per year increases the effect of compounding levels off or reaches a limit. Therefore the natural exponential e can be substituted into the compound interest formula to accomplish **continuous compounding.**

How to

Find the accumulated amount (future value) for continuous compounding:

1. Determine the principal (P), rate per year (r), and the number of years (t).
2. Evaluate the formula

   $A = Pe^{rt}$

   where A = accumulated amount or future value

   P = principal

   r = rate per year

   t = time in years

EXAMPLE To find the compound amount of $5,000 invested at an annual rate of 4% compounded continuously for 5 years:

$P = 5,000, r = 0.04$ (from 4%), $t = 5$

$A = Pe^{rt}$	Substitute values.
$A = 5,000 \ e^{(0.04)(5)}$	Simplify exponents.
$A = 5,000e^{0.20}$	Raise to the power.
$A = 5,000(1.221402758)$	Multiply.
$A = 6,107.01$	

The accumulated amount is $6,107.01.

An **annuity** is a fund that accumulates compound interest as periodic payments add to the principal. An **ordinary annuity** has periodic payments that are made at the end of each payment period. An **annuity due** has periodic payments that are made at the beginning of each payment period. The difference in the two types of annuities is that the annuity due earns interest on the first period and the ordinary annuity does not.

▲ APPLICATION: Future Value of an Ordinary Annuity

How to

Find the future value of an ordinary annuity:

Apply the formula

$$FV = P\left[\frac{(1 + R)^N - 1}{R}\right]$$

where FV = future value of an ordinary annuity

P = amount of the periodic payment

R = interest rate per period

N = total number of periods

EXAMPLE   To find the future value of an ordinary annuity of $6,000 for five years at 6% annual interest compounded semiannually:

$$FV = P\left[\frac{(1 + R)^N - 1}{R}\right]; P = \$6,000; R = \frac{0.06}{2} = 0.03; N = 5(2) = 10$$

$$FV = 6,000\left[\frac{(1 + 0.03)^{10} - 1}{0.03}\right] \qquad \text{Simplify innermost grouping.}$$

$$FV = 6,000\left[\frac{(1.03)^{10} - 1}{0.03}\right] \qquad \text{Raise to the power.}$$

$$FV = 6,000\left[\frac{1.343916379 - 1}{0.03}\right] \qquad \text{Subtract in numerator.}$$

$$FV = 6,000\left[\frac{0.343916379}{0.03}\right] \qquad \text{Divide in grouping.}$$

$$FV = 6,000(11.46387931) \qquad \text{Multiply.}$$

$$FV = \$68,783.28$$

▲ APPLICATION: Future Value of an Annuity Due

How to

Find the future value of an annuity due:

Apply the formula

$$FV = P\left[\frac{(1 + R)^N - 1}{R}\right](1 + R)$$

where FV = future value of an annuity due

P = amount of the periodic payment

R = interest rate per period

N = total number of periods

EXAMPLE   To find the future value of a quarterly annuity due of $100 for five years at 5% compounded quarterly:

$$FV = P\left[\frac{(1 + R)^N - 1}{R}\right][1 + R]$$

$$P = \$100, R = \frac{0.05}{4} = 0.0125, N = 5(4) = 20$$

$$FV = 100\left[\frac{(1 + 0.0125)^{20} - 1}{0.0125}\right][1 + 0.0125]$$

$$FV = 100\left[\frac{(1.0125)^{20} - 1}{0.0125}\right][1.0125]$$

$$FV = 100\left[\frac{1.282037232 - 1}{0.0125}\right][1.0125]$$

$$FV = 100\left[\frac{0.282037232}{0.0125}\right][1.0125]$$

$$FV = 100(22.56297854)(1.0125)$$

$$FV = \$2,284.50$$

● **RELATED TOPIC: Future value of an investment**

The are occasions when you would like to equate an ordinary annuity with a lump sum investment. That is, how much money would need to be invested today to accumulate to the same amount as a periodic payment over a specified time at a specified rate. This lump sum equivalent is the **present value of an ordinary annuity.**

Exponential
Logarithmic

▲ APPLICATION: Present Value of an Annuity

How to

Find the present value of an ordinary annuity:

Apply the formula

$$PV = P\left[\frac{(1 + R)^N - 1}{R(1 + R)^N}\right]$$

where PV = present value of an ordinary annuity (lump sum equivalent)

P = amount of the periodic payment

R = interest rate per period

N = total number of periods

EXAMPLE   Johanna Helba could establish an ordinary annuity by depositing $1,600 at the end of each year for 10 years at 6% annual interest. Or she could set aside a lump sum today. To find the amount she needs to set aside today at the same rate, 6%, so that she has the same as she would have accumulated with the annuity:

$PV = P\left[\dfrac{(1 + R)^N - 1}{R(1 + R)^N}\right]$ Substitute known values.

$P = 1,600; R = \dfrac{0.06}{1} = 0.06; N = 10(1) = 10$

$PV = 1,600\left[\dfrac{(1 + 0.06)^{10} - 1}{0.06(1 + 0.06)^{10}}\right]$ Evaluate innermost groupings.

$PV = 1,600\left[\dfrac{(1.06)^{10} - 1}{0.06(1.06)^{10}}\right]$ Evaluate numerator and denominator.

$PV = 1,600\left[\dfrac{0.7908476965}{0.1074508618}\right]$ Divide in grouping and multiply.

$PV = \$11,776.14$

When you have a specific future goal or target amount that you want to accumulate, a **sinking fund payment** is the amount you would invest in periodic payments to reach this goal. To determine the sinking fund payment the *known values* are the future goal or amount, the amount of time, and the expected or guaranteed interest rate.

▲ **APPLICATION: Sinking Fund Payment Needed for Specific Future Value**

How to

Find the sinking fund payment to produce a specified future value:

Apply the formula

$$P = FV\left[\frac{R}{(1 + R)^N - 1}\right]$$

where P = sinking fund payment

FV = future value or goal

R = interest rate per period

N = total number of periods

Exponential Logarithmic

EXAMPLE   A professional football team has established a sinking fund to retire a bond issue of $500,000, which is due in 10 years. The account pays 8% quarterly interest. To find the amount of the sinking fund payment:

$$P = FV\left[\frac{R}{(1 + R)^N - 1}\right]$$ Substitute known values.

$$FV = \$500,000; R = \frac{0.08}{4} = 0.02; N = 10(4) = 40$$

$$P = 500,000\left[\frac{0.02}{(1 + 0.02)^{40} - 1}\right]$$ Simplify grouping in denominator.

$$P = 500,000\left[\frac{0.02}{(1.02)^{40} - 1}\right]$$

$$P = 500,000\left[\frac{0.02}{2.208039664 - 1}\right]$$

$$P = 500,000\left[\frac{0.02}{1.208039664}\right]$$ Divide.

$$P = 500,000[0.0165557478]$$ Multiply.

$$P = \$8,277.87$$

▲ **APPLICATION: How to Become a Millionaire**

EXAMPLE   Suppose that at age 25 you decide that you want to have one million dollars in an investment fund at age 65. How much should you invest each month at 10% annual interest compounded monthly to reach your goal?

To find how much of the one million dollars you will have paid:

Use the formula to find a sinking fund payment. $FV = \$1,000,000$; $R = \frac{0.10}{12} = 0.0083$ per month; $N = 40$ years $\times$ 12 months per year $= 480$ payments.

$$P = FV\left[\frac{R}{(1 + R)^N - 1}\right]$$ Substitute known values.

$$P = \$1,000,000\left[\frac{0.0083}{(1 + 0.0083)^{480} - 1}\right]$$ Simplify grouping in denominator.

$$P = \$1,000,000\left[\frac{0.0083}{1.0083^{480} - 1}\right]$$ Evaluate denominator.

$$P = \$1,000,000\left[\frac{0.0083}{51.85526454}\right]$$ Divide in grouping and multiply.

$$P = \$160.06$$ Monthly payment.

$$\$160.06 \times 480 = \$76,828.80$$ Total of all payments or total amount of investment.

TIP

How does rounding affect a series of calculations?

The answer in the previous example will vary if the rate per period is rounded to a different number of decimal places. Built-in computer or calculator functions will calculate with more decimal places and yield a more accurate answer.

Finding the monthly payment to repay a loan is similar to the process for finding the sinking fund payment. The repayment of the loan in equal installments that are applied to the principal and interest over a specified amount of time is called the **amortization of a loan.**

▲ APPLICATION: Monthly Payment for an Amortized Loan

How to

Find the monthly payment for an amortized loan:

Apply the formula $M = P\left[\dfrac{R}{1 - (1 + R)^{-N}}\right]$
where M = monthly payment; P = principal or initial amount of the loan; R = interest rate per month; N = total number of months

EXAMPLE   To find the monthly payment on a 25-year home mortgage of $135,900 at 8%:

$$P = \$135{,}900;\ R = \frac{0.08}{12} = 0.00667;\ N = 25(12) = 300$$

$$M = P\left[\frac{R}{1 - (1 + R)^{-N}}\right] \qquad \text{Substitute values.}$$

$$M = 135{,}900\left[\frac{0.00667}{1 - (1 + 0.00667)^{-300}}\right] \qquad \text{Simplify denominator.}$$

$$M = 135{,}900\left[\frac{0.00667}{1 - (1.00667)^{-300}}\right]$$

$$M = 135{,}900\left[\frac{0.00667}{1 - 0.1361012491}\right]$$

$$M = 135{,}900\left[\frac{0.00667}{0.8638987509}\right] \qquad \begin{array}{l}\text{Divide in grouping and}\\ \text{multiply.}\end{array}$$

$$M = \$1{,}049.26$$

■ 17–2 Logarithmic Expressions and Equations

The three basic components of a power are the base, exponent, and result of exponentiation or power. The word *power* has more than one meaning, so we use the **result of exponentiation** terminology.

$$\overset{\text{exponent}}{\underset{\text{base}}{3^{4}}} = 81 \;\leftarrow\; \text{result of exponentiation (power)}$$

Exponential · Logarithmic

When finding a power or the result of exponentiation, you are given the base and exponent. When finding a root, you know the radicand (base) and the index of the root (exponent). A third type of calculation involves finding the exponent when the base and the result of exponentiation are given. The exponent in this process is called the **logarithm.** In the logarithmic form $\log_b x = y$, b is the **base,** y is the **exponent or logarithm,** and x is the **result of exponentiation.** This equation is read, "The log of x to the base b is y." Logarithm is written in abbreviated form as "log." In the exponential form $x = b^y$, b is also the base, y is the exponent, and x is the result of the exponentiation.

Algebraic expressions written in logarithmic or exponential form have the same three components: base, exponent, result of exponentiation. Let's examine the mapping from one form to the other.

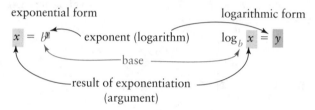

exponential form   logarithmic form

$x = b^y$ exponent (logarithm) $\log_b x = y$
                 base
         result of exponentiation
                (argument)

How to

Convert an exponential equation to a logarithmic equation:

If $x = b^y$, then $\log_b x = y$, provided that $b > 0$ and $b \neq 1$.

1. The exponent in the exponential form is the variable solved for in the logarithmic form.
2. The base in the exponential form is the base in the logarithmic form.
3. The variable solved for in the exponential form is the result of exponentiation in logarithmic form. This result is sometimes referred to as the **argument.**

EXAMPLE   To change from exponential to logarithmic form:

    (a) $2^4 = 16$ (b) $3^2 = 9$ (c) $2^{-2} = \dfrac{1}{4}$

(a) $2^4 = 16$ converts to $\log_2 16 = 4$ Base = 2, exponent = 4, result of exponentiation = 16.

(b) $3^2 = 9$ converts to $\log_3 9 = 2$ Base = 3, exponent = 2, result of exponentiation = 9.

(c) $2^{-2} = \dfrac{1}{4}$ converts to $\log_2 \dfrac{1}{4} = -2$ Base = 2, exponent = −2, result of exponentiation = $\frac{1}{4}$.

● **RELATED TOPIC: Negative exponents**

How to

Convert a logarithmic equation to an exponential equation:

$\log_b x = y$ converts to $x = b^y$, provided that $b > 0$ and $b \neq 1$.

1. The dependent variable (variable solved for) in the logarithmic form is the exponent in the exponential form.
2. The base in the logarithmic form is the base in the exponential form.
3. The result of the exponentiation in logarithmic form is the dependent variable (variable solved for) in exponential form.

EXAMPLE   Write in exponential form.

(a) $\log_2 32 = 5$ (b) $\log_3 81 = 4$

(c) $\log_5 \dfrac{1}{25} = -2$ (d) $\log_{10} 0.001 = -3$

(a) $\log_2 32 = 5$ converts to $2^5 = 32$ Base = 2, exponent = 5, result of exponentiation = 32.

(b) $\log_3 81 = 4$ converts to $3^4 = 81$ Base = 3, exponent = 4, result of exponentiation = 81.

(c) $\log_5 \dfrac{1}{25} = -2$ converts to $5^{-2} = \dfrac{1}{25}$ Base = 5, exponent = −2, result of exponentiation = $\frac{1}{25}$.

(d) $\log_{10} 0.001 = -3$ converts to $10^{-3} = 0.001$

Base = 10, exponent = −3, result of exponentiation = 0.001.

When the base of a logarithm is 10, the logarithm is referred to as a **common logarithm**. If the base is omitted in a logarithmic expression, the base is

(side margin) Exponential Logarithmic

understood to be 10. Thus, $\log_{10} 1{,}000 = 3$ is normally written as $\log 1{,}000 = 3$. On a calculator, expressions containing common logarithms can be evaluated using the $\boxed{\log}$ key.

A logarithm with a base of e is a **natural logarithm** and is abbreviated as *ln*. Calculators normally have a $\boxed{\ln}$ key. Thus, $\log_e 1 = 0$ is normally written as $\ln 1 = 0$. Expressions containing natural logarithms can be evaluated using the $\boxed{\ln}$ key.

Calculators are used regularly when working with logarithms.

How to

Evaluate a logarithm with a base b other than 10 or e:

$$\log_b a = \frac{\log a}{\log b}$$

A similar process can be used with natural logarithms.

$$\log_b a = \frac{\ln a}{\ln b}$$

To check, evaluate b^x where b is the base of the logarithm and x is the logarithm. The result should equal the argument.

EXAMPLE   Find $\log_7 343$.

$$\log_7 343 = \frac{\log 343}{\log 7} = 3 \qquad \boxed{\log}\ 343\ \boxed{\div}\ \boxed{\log}\ 7\ \boxed{=}\ \Rightarrow 3;\ \text{Check: } 7^3 = 343.$$

▲ **APPLICATION: How Long Does It Take an Investment to Reach a Specified Amount?**

How to

Find the time for an investment to reach a specified amount if compounded continuously:

$$t = \frac{(\ln A - \ln P)}{r}$$

t = time (in years) of investment

A = accumulated amount

P = initial invested amount

r = annual rate of interest

EXAMPLE   To find the length of time for $32,750 to reach $47,650.97 when invested at 5% compounded continuously:

$A = \$47,650.97, P = \$32,750, r = 0.05$

$$t = \frac{\ln A - \ln P}{r}$$

$$t = \frac{\ln 47,650.97 - \ln 32,750}{0.05}$$

$$t = \frac{10.77165827 - 10.39665824}{0.05}$$

$$t = \frac{0.3750000246}{0.05}$$

$t = \textbf{7.5 years}$

∟

▲ APPLICATION: How Long Does It Take an Investment to Double?

EXAMPLE   How long does it take an investment of $10,000 to double if it is invested at 5% annual interest compounded continuously?

$A = \$20,000; P = \$10,000; r = 0.05$

$$t = \frac{\ln A - \ln P}{r}$$

$$t = \frac{\ln 20,000 - \ln 10,000}{0.05}$$

$$t = \frac{9.903487553 - 9.210340372}{0.05}$$

$$t = \frac{0.6931471806}{0.05}$$

$t = \textbf{13.9 years}$

∟

● **RELATED TOPICS: Compound amount, Compound interest**

Exponential
Logarithmic

How long does it take an investment (lump sum) to double or "The Rule of 72"

A good "rule of thumb" to find the length of time it takes an investment to double if it is compounded at least annually is to divide 72 by the annual interest rate.

For example, if a sum is invested at 5% (see previous Example) it should double in approximately 14 years.

$$\frac{72}{5} = 14.4$$

Variable interpretations change from formula to formula

In the financial formulas we have examined the interpretation of a variable may change from formula to formula. For example, P may represent the principal or initial amount in one formula and the periodic payment in another formula.

Interpreting the variables appropriately is a critical part of evaluating any formula.

Measurement and Geometry

CHAPTER 18

Direct Measurement

■ 18–1 The U.S. Customary System of Measurement

The **U.S. customary system of measurement** continues to be used in the United States for many applications. The most common units of measure are:

U.S. Customary Units of Length or Distance

12 inches (in.)[a] = 1 foot (ft)[b]	36 inches (in.) = 1 yard (yd)
3 feet (ft) = 1 yard (yd)	5,280 feet (ft) = 1 mile (mi)

[a]The symbol ″ means inches (8″ = 8 in.) or seconds (60″ = 60 seconds).
[b]The symbol ′ means feet (3′ = 3 ft) or minutes (60′ = 60 minutes).

U.S Customary Units of Weight or Mass

16 ounces (oz) = 1 pound (lb)
2,000 pounds (lb) = 1 ton (T)

U.S. Customary Units of Liquid Capacity or Volume

3 teaspoons (t) = 1 tablespoon (T)   2 tablespoons (T) = 1 ounce (oz)
8 ounces (oz) = 1 cup (c)   4 cups (c) = 1 quart (qt)
2 cups (c) = 1 pint (pt)   4 quarts (qt) = 1 gallon (gal)
2 pints (pt) = 1 quart (qt)

Using the relationship between two units of measure, we can form a ratio that has a value of 1 in two different ways. We call this type of ratio a **unity ratio**. A *unity ratio* is a ratio of measures that has a value of 1. Every relationship between two measures can be used to form two different unity ratios.

One foot = 12 inches

$$\frac{1 \text{ ft}}{12 \text{ in.}} \qquad \frac{12 \text{ in.}}{1 \text{ ft}}$$

In each unity ratio, the value of the numerator equals the value of the denominator. A ratio with the numerator and denominator equal, has a value of 1.

How to

Change from one U.S. customary unit of measure to another using unity ratios:

1. Set up the original amount as a fraction with the original unit of measure in the numerator.
2. Multiply this by a unity ratio with the original unit in the denominator and the new unit in the numerator.
3. Reduce like units of measure and all numbers wherever possible.

EXAMPLE   To find the number of inches in 5 ft:
Multiply 5 ft by a unity ratio that contains both inches and feet.
    Because 5 ft is a whole number, we write it with 1 as the denominator.

$$\frac{5 \text{ ft}}{1} \left(\frac{}{} \right)$$ Place the original unit with the 5 in the *numerator* of the first fraction.

$$\frac{5 \text{ ft}}{1} \left(\frac{}{\text{ft}} \right)$$ We are changing *from* (feet), so place ft in the *denominator* of the unity ratio, which is shown in parentheses. This allows us to reduce the units later.

$$\frac{5 \text{ ft}}{1} \left(\frac{\text{in.}}{\text{ft}} \right)$$

To change *to* inches, place inches in the *numerator* of the unity ratio.

$$\frac{5 \text{ ft}}{1} \left(\frac{12 \text{ in.}}{1 \text{ ft}} \right) = 60 \text{ in.}$$

Place in the unity ratio the numerical values that make these two units of measure equivalent (1 ft = 12 in.). Complete the calculation, reducing wherever possible.

5 ft = 60 in.

EXAMPLE   To find the number of ounces in 2 gallons:
Change from gallons to ounces using the following conversions:

gallons → quarts → pints → cups → ounces

$$\frac{2 \text{ gal}}{1} \left(\frac{\text{qt}}{\text{gal}} \right) \left(\frac{\text{pt}}{\text{qt}} \right) \left(\frac{\text{c}}{\text{pt}} \right) \left(\frac{\text{oz}}{\text{c}} \right) =$$

Set up ratios and insert relationship numbers.

$$\frac{2 \text{ gal}}{1} \left(\frac{4 \text{ qt}}{1 \text{ gal}} \right) \left(\frac{2 \text{ pt}}{1 \text{ qt}} \right) \left(\frac{2 \text{ c}}{1 \text{ pt}} \right) \left(\frac{8 \text{ oz}}{1 \text{ c}} \right) =$$

Reduce measures.

$$2(4)(2)(2)(8) \text{ oz} = 256 \text{ oz}$$ Multiply.

2 gal = 256 oz.

Alternative method   We can work this same problem with fewer unity ratios with some preliminary calculations using the relationships. Multiply 4 (cups) by 8 (ounces per cup) to find the number of ounces in a quart. So 32 oz = 1 qt.

$$\frac{2 \text{ gal}}{1} \left(\frac{4 \text{ qt}}{1 \text{ gal}} \right) \left(\frac{32 \text{ oz}}{1 \text{ qt}} \right)$$

Set up unity ratios.

$$\frac{2 \text{ gal}}{1} \left(\frac{4 \text{ qt}}{1 \text{ gal}} \right) \left(\frac{32 \text{ oz}}{1 \text{ qt}} \right) = 256 \text{ oz}$$

Reduce and multiply.

2 gal = 256 oz.

Estimation and dimension analysis

When estimating unit conversions, see if the new unit is larger or smaller than the original unit.

Larger to smaller

Each larger unit can be divided into smaller units. Larger-to-smaller conversions mean *more* smaller units. *More* implies multiplication.

To convert a U.S. customary unit to a desired *smaller* unit, *multiply* the number of larger units by the number of smaller units that equals 1 larger unit.

2 yd = _____ ft   **The smaller unit is feet: 3 ft = 1 yd.**

2 × 3 ft = 6 ft   **Multiply number of yards by 3 ft.**

2 yd = 6 ft.   **Dimension analysis:** $\dfrac{2 \text{ yd}}{1} \times \dfrac{3 \text{ ft}}{1 \text{ yd}} = 6 \text{ ft}$

The key word clues are:

Larger to smaller unit → obtain more units → multiply

Smaller to larger

Several small units combine to make one large unit. Thus, smaller-to-larger conversions mean *fewer* large units. *Fewer* implies division.

To convert a U.S. customary unit to a *larger* unit, *divide* the original unit by the number of smaller units that equals 1 desired larger unit.

12 ft = _____ yd   **The larger unit is yards: 1 yd = 3 ft.**

12 ÷ 3 = 4   **Divide by 3 ft to get yards.**

12 ft = 4 yd   **Dimension analysis:** $\dfrac{12 \text{ ft}}{1} \times \dfrac{1 \text{ yd}}{3 \text{ ft}} = 4 \text{ yd}$

Using key word clues:

Smaller to larger unit → obtain fewer units → divide

Estimation can catch errors in setting up the problem, but it is not as likely to catch calculation errors.

Unity ratios can be used to develop conversion factors. With conversion factors you always multiply to change from one measuring unit to another.

How to

Develop a conversion factor for converting from one measure to another:

1. Write a unity ratio that changes the given unit to the new unit.
2. Change the fraction (or ratio) to its decimal equivalent by dividing the numerator by the denominator.

EXAMPLE   To develop two conversion factors relating pounds and ounces:

Pounds to Ounces

$$\frac{pounds}{1} \left(\frac{ounces}{pounds} \right)$$

$$\frac{pounds}{1} \left(\frac{16\ ounces}{1\ pound} \right)$$

$$\frac{16}{1} = 16$$

pounds × 16 = ounces

Ounces to Pounds

$$\frac{ounces}{1} \left(\frac{pounds}{ounces} \right)$$

$$\frac{ounces}{1} \left(\frac{1\ pound}{16\ ounces} \right)$$

$$1 \div 16 = 0.0625$$

ounces × 0.0625 = pounds

How to

Change from one U.S. customary unit of measure to another using conversion factors:

1. Select the appropriate conversion factor.
2. Multiply the original measure by the conversion factor.

EXAMPLE   To use a conversion factor to convert 54 ounces to pounds:

ounces × 0.0625 = pounds   Conversion factor for ounces to pounds.

56 × 0.0625 = 3.5 pounds   Multiply.

56 ounces is 3.5 pounds.

U.S. Customary Conversion Factors

		TO CHANGE	
	From	To	Multiply By
Length or Distance			
12 inches (in.) = 1 foot (ft)	feet	inches	12
	inches	feet	0.0833333
3 feet (ft) = 1 yard (yd)	yards	feet	3
	feet	yards	0.3333333
36 inches (in.) = 1 yard (yd)	yards	inches	36
	inches	yards	0.0277778
5,280 feet (ft) = 1 mile (mi)	miles	feet	5280
	feet	miles	0.0001894
Weight or Mass			
16 ounces (oz) = 1 pound (lb)	pounds	ounces	16
	ounces	pounds	0.0625
2,000 pounds (lb) = 1 ton (T)	pounds	tons	2000
	tons	pounds	0.0005
Liquid Capacity or Volume			
8 ounces (oz) = 1 cup (c)	cups	ounces	8
	ounces	cups	0.125
2 cups (c) = 1 pint (pt)	cups	pints	2
	pints	cups	0.5
2 pints (pt) = 1 quart (qt)	quarts	pints	2
	pints	quarts	0.5
4 quarts (qt) = 1 gallon (gal)	gallons	quarts	4
	quarts	gallons	0.25

Additional conversion factors are found in Appendix A–6.

Direct Measurement

■ 18–2 Basic Operations with U.S. Customary Measures

Measures that use two or more units are called **mixed measures.** A mixed measure is in **standard notation** if the number associated with each unit of measure is smaller than the number required to convert to the next larger unit. The number in the largest unit of measure given may or may not be converted as desired.

How to

Express mixed measures in standard notation:

1. Start with the smallest unit of measure and determine if there are enough units to make one or more of the next unit.
2. Regroup to make as many of the larger units as possible.
3. Add the new larger unit to the original units.
4. Repeat the process with each given measuring unit.

EXAMPLE   To express 2 yd 4 ft 16 in. in standard notation:

2 yd 4 ft 16 in. = 2 yd 4 ft 12 in. + 4 in. = 2 yd 4 ft + 1 ft + 4 in.
Examine inches.

= 2 yd 5 ft 4 in. = 2 yd 3 ft + 2 ft 4 in.
Examine feet.

= 2 yd + 1 yd 2 ft 4 in. = **3 yd 2 ft 4 in.**
└ Standard notation.

Standard conventions

We can say that 3 ft 15 in. is 4 ft 3 in. in standard notation. Why not change 4 ft 3 in. to 1 yd 1 ft 3 in.? There may be situations when 1 yd 1 ft 3 in. is the desirable form; however, in general, we keep the same units of measure in standard notation. 3 qt 3 pt = 4 qt 1 pt (instead of 1 gal 1 pt).

How to

Add unlike U.S. customary measures:

1. Convert to a common U.S. customary unit.
2. Add.

EXAMPLE   To add 3 ft + 2 in.:

Because 2 in. is a fraction of a foot, we can avoid working with fractions by converting the larger unit (feet) to the smaller unit (inches).

$$3 \text{ ft} = \frac{3 \cancel{\text{ft}}}{1} \left(\frac{12 \text{ in.}}{1 \cancel{\text{ft}}} \right) = 36 \text{ in.}$$

$$\boxed{3 \text{ ft}} + 2 \text{ in.} = \qquad 3 \text{ ft} = 36 \text{ in.}$$

$$\boxed{36 \text{ in.}} + 2 \text{ in.} = 38 \text{ in.}$$

How to

Add mixed U.S. customary measures:

1. Align the measures vertically so the common units are written in the same vertical column.
2. Add.
3. Express the sum in standard notation.

EXAMPLE   To add 6 lb 7 oz and 3 lb 13 oz and write the answer in standard form:

$$
\begin{array}{r}
6 \text{ lb } \;\; 7 \text{ oz} \\
+ \; 3 \text{ lb } 13 \text{ oz} \\
\hline
9 \text{ lb } 20 \text{ oz} \quad \text{20 oz = 1 lb 4 oz}
\end{array}
$$

9 lb 20 oz = 10 lb 4 oz

How to

Subtract unlike U.S. customary units:

1. Convert to a common U.S. customary unit.
2. Subtract.

EXAMPLE   To subtract 15 in. from 2 ft:

$$2 \text{ ft} = 24 \text{ in.}$$

24 in. − 15 in. = **9 in.**   Changing 15 in. to feet gives us a mixed number, so it is easier to convert 2 ft to inches.

How to

Subtract mixed U.S. customary measures:

1. Align the measures vertically so that the common units are written in the same vertical column.
2. Subtract.

EXAMPLE To subtract 5 ft 3 in. from 7 ft 4 in.:

$$
\begin{array}{r}
7 \text{ ft } 4 \text{ in.} \\
- 5 \text{ ft } 3 \text{ in.} \\
\hline
2 \text{ ft } 1 \text{ in.}
\end{array}
$$
Align like measures in a vertical line, then subtract.

Sometimes when we subtract mixed measures, we subtract a larger unit from a smaller one. In this case, we use our knowledge of borrowing.

How to

Subtract a larger U.S. customary unit from a smaller unit in mixed measures:

1. Align the common measures in vertical columns.
2. Borrow one unit from the next larger unit of measure in the minuend, convert to the equivalent smaller unit, and add it to the smaller unit.
3. Subtract the measures.

EXAMPLE To subtract 3 lb 12 oz from 7 lb 8 oz:

$$
\begin{array}{r}
7 \text{ lb } 8 \text{ oz} \\
- 3 \text{ lb } 12 \text{ oz}
\end{array}
$$

We always begin subtraction with the smallest unit, which should be on the *right*. In this example, notice that 12 oz cannot be subtracted from 8 oz, so we rewrite 7 lb as 6 lb 16 oz. Then, we add 16 oz to 8 oz to get 24 oz.

$$
\begin{array}{r}
7 \text{ lb } 8 \text{ oz} = 6 \text{ lb } 16 \text{ oz} + 8 \text{ oz} = \quad 6 \text{ lb } 24 \text{ oz} \\
- 3 \text{ lb } 12 \text{ oz} = \quad\quad\quad\quad\quad\quad\quad\quad - 3 \text{ lb } 12 \text{ oz} \\
\hline
3 \text{ lb } 12 \text{ oz}
\end{array}
$$

How to

Multiply a U.S. customary measure by a number:

1. Multiply the numbers associated with each unit of measure by the given number.
2. Write the resulting measure in standard notation.

EXAMPLE   A container holds 21 gal 3 qt of weed killer. To determine how much eight containers hold:

$$
\begin{array}{r}
21 \text{ gal} \quad 3 \text{ qt} \\
\times \qquad\qquad 8 \\
\hline
168 \text{ gal } 24 \text{ qt} \\
\end{array}
\quad 24 \text{ qt} = 6 \text{ gal}
$$

168 gal 24 qt = 168 gal + 6 gal = 174 gal in standard notation

The eight containers hold 174 gal.

Length measures multiplied by like length measures produce square units. Square measures indicate areas.

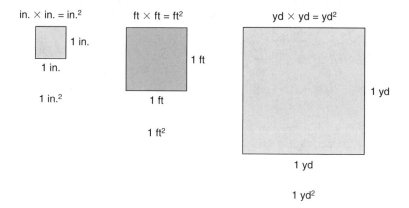

Area gives us a different perspective on units of measure. Units of area measure surfaces instead of lengths.

Length × Length = Length^2 or Area

How to

Multiply a length measure by a like length measure:

1. Multiply the numbers associated with each like unit of measure.
2. Write the product as a square unit of measure.

EXAMPLE A desktop is 2 ft × 3 ft. To determine the number of square feet in the surface:

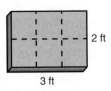

2 ft

3 ft

2 ft × 3 ft = **6 ft²**

How to

Divide a U.S. customary measure by a number that divides evenly into each measure:

1. Divide the numbers associated with each unit of measure by the given number.
2. Write the resulting measure in standard notation.

EXAMPLE To determine how much milk is needed for a half-recipe if the original recipe calls for 2 gal 2 qt:

$2\overline{)2 \text{ gal } 2 \text{ qt}}$ Divide each measure by 2.

$\dfrac{1 \text{ gal } 1 \text{ qt}}{2\overline{)2 \text{ gal } 2 \text{ qt}}}$

The half-recipe requires 1 gal 1 qt of milk.

How to

Divide a U.S. customary measure by a number that does not divide evenly into each measure:

1. Set up the problem and proceed as in long division.
2. When a remainder occurs after subtraction, convert the remainder to the unit of measure used in the next smaller measure and add it to the quantity in the next smaller measure.
3. Divide the given number into this next smaller unit.
4. If a remainder occurs when the smallest unit is divided, express the remainder as a fractional part of the smallest unit.

EXAMPLE   To divide 5 gal 3 qt 1 pt by 3:

$$\begin{array}{r} 1\text{ gal} \quad 3\text{ qt} \quad 1\frac{2}{3}\text{ pt} \\ 3\overline{)5\text{ gal} \quad 3\text{ qt} \quad 1\text{ pt}} \\ \underline{3\text{ gal}} \\ 2\text{ gal} = \underline{8\text{ qt}} \\ 11\text{ qt} \\ \underline{9\text{ qt}} \\ 2\text{ qt} = 4\text{ pt} \\ 5\text{ pt} \\ \underline{3\text{ pt}} \\ 2\text{ pt} \end{array}$$

5 gal ÷ 3 = 1 gal, remainder 2 gal
2 gal = 8 qt
3 qt + 8 qt = 11 qt
11 qt ÷ 3 = 3 qt, remainder 2 qt
2 qt = 4 pt
1 pt + 4 pt = 5 pt
5 pt ÷ 3 = 1 pt, remainder 2 pt

Write the final remainder, 2, as a fraction $\frac{2}{3}$, and add to 1 pint to get $1\frac{2}{3}$ pints.

Thus, 1 gal 3 qt $1\frac{2}{3}$ pt is the solution.

How to

Divide a U.S. customary measure by a U.S. customary measure:

1. Convert both measures to the same unit if they are different.
2. Write the division as a fraction, including the common unit in the numerator and the denominator.
3. Reduce the units and divide the numbers.

EXAMPLE To divide 7 ft 6 in. by 10 in.:

$$7 \text{ ft} = \frac{7 \text{ ft}}{1} \left(\frac{12 \text{ in.}}{1 \text{ ft}} \right) = 84 \text{ in.} \qquad \text{Convert 7 ft 6 in. to inches.}$$

7 ft 6 in. = 84 in. + 6 in. = 90 in.

90 in. ÷ 10 in. Write this division in fraction form. Divide to see that the common units reduce. The answer is a number (not a measure) telling *how many times* one measure divides into another.

$$\frac{90 \text{ in.}}{10 \text{ in.}} = 9 \text{ pieces}$$

There are 9 pieces, with 10 inches per piece.

A **rate measure** is a ratio of two different kinds of measures. It is often referred to as a *rate*. Some examples of rates are 55 miles per hour, 20 cents per mile, and 3 gallons per minute. In each of these rates, the word *per* means *divided by*.

The rate 55 miles per hour means 55 miles ÷ 1 hour or $\frac{55 \text{ mi}}{1 \text{ hr}}$. The rate 20 cents per mile means 20 cents ÷ 1 mile or $\frac{20 \text{ cents}}{1 \text{ mi}}$. The rate 3 gallons per minute means 3 gallons ÷ 1 minute or $\frac{3 \text{ gal}}{1 \text{ min}}$.

Many rate measures involve measures of time. The units used to measure time are universally accepted and used around the world. The basic units of time are the year, month, week, day, hour, minute, and second.

Units of Time

1 year (yr) = 12 months (mo)	1 day (da) = 24 hours (hr)
1 year (yr) = 365 days (da)	1 hour (hr) = 60 minutes (min)[a]
1 week (wk) = 7 days (da)	1 minute (min) = 60 seconds (sec)[b]

[a]The symbol ′ means feet (3′ = 3 ft) or minutes (60′ = 60 minutes).
[b]The symbol ″ means inches (8″ = 8 in.) or seconds (60″ = 60 seconds).

How to

Convert one U.S. customary rate measure to another:

1. Compare the units of both numerators and both denominators to determine which units change.

2. Multiply each unit that changes by a unity ratio containing the new unit so that the unit to be changed will reduce.

EXAMPLE   To change $8\dfrac{\text{pt}}{\text{min}}$ to $\dfrac{\text{qt}}{\text{min}}$:

Examine the rates:

Numerators—pints to quarts
Denominators—no change

$$\dfrac{\text{pt}}{\text{min}}\left(\dfrac{\text{qt}}{\text{pt}}\right) = \dfrac{\text{qt}}{\text{min}} \qquad \text{Develop an appropriate unity ratio.}$$

Estimation   Pints to quarts is *smaller* to *larger,* so there will be fewer quarts.

$$\dfrac{\overset{4}{8}\,\text{pt}}{\text{min}}\left(\dfrac{1\,\text{qt}}{\underset{1}{2}\,\text{pt}}\right) = \dfrac{4\,\text{qt}}{\text{min}} \qquad \text{Multiply.}$$

Interpretation $8\,\dfrac{\text{pt}}{\text{min}}$ **equals** $4\,\dfrac{\text{qt}}{\text{min}}.$

⌞

▲ **APPLICATION: Changing Miles per Hour to Feet per Second**

EXAMPLE   To change 60 miles per hour to feet per second:

$$60 \text{ miles per hour} = 60\,\dfrac{\text{mi}}{\text{hr}} \qquad \text{Write rate as a fraction.}$$

$$\text{feet per second} = \dfrac{\text{ft}}{\text{sec}}$$

Numerators—miles to feet   Examine the changes in the measures.
Denominators—hours to seconds

$$60\,\dfrac{\text{mi}}{\text{hr}}\overset{\substack{\text{(miles}\\\text{to}\\\text{feet)}}}{\left(\dfrac{5280\,\text{ft}}{1\,\text{mi}}\right)}\overset{\substack{\text{(hours}\\\text{to}\\\text{minutes)}}}{\left(\dfrac{1\,\text{hr}}{60\,\text{min}}\right)}\overset{\substack{\text{(minutes}\\\text{to}\\\text{seconds)}}}{\left(\dfrac{1\,\text{min}}{60\,\text{sec}}\right)} \qquad \begin{array}{l}\text{Develop appropriate unity ratios}\\\text{and reduce.}\end{array}$$

$$\dfrac{5280\,\text{ft}}{60\,\text{sec}} = 88\,\dfrac{\text{ft}}{\text{sec}} \qquad \text{Divide.}$$

At 60 miles per hour you are traveling 88 feet per second.

⌞

■ 18–3 The Metric System of Measurement

The **metric system** is an international system of measurement that uses standard units and prefixes to indicate powers of 10.

In the metric system, or the International System of Units (SI), there is a standard unit for each type of measurement. The **meter** is used for length or distance, the **gram** is used for weight or mass, and the **liter** is used for capacity or volume. A series of prefixes are affixed to the standard unit to indicate measures greater than the standard unit or less than the standard unit.

The most common metric prefixes for units *smaller* than the standard unit are:

$$\textbf{deci-}\ \frac{1}{10}\ \text{of} \qquad \textbf{centi-}\ \frac{1}{100}\ \text{of} \qquad \textbf{milli-}\ \frac{1}{1,000}\ \text{of}$$

The most common prefixes for units *larger* than the standard unit are

deka- 10 times   **hecto-** 100 times   **kilo-** 1,000 times

Other prefixes for larger and smaller measures are given on p. 301 and in Appendix A–7.

We can relate the prefixes to our decimal system. The standard unit (whether meter, gram, or liter) corresponds to the **ones** place. All the places to the left are powers of the standard unit. That is, the value of **deka-** (some references use *deca-*) is 10 times the standard unit, the value of **hecto-** is 100 times this unit, the value of **kilo-** is 1,000 times this unit, and so on. All the places to the right of the standard unit are subdivisions of the standard unit. That is, the value of **deci-** is $\frac{1}{10}$ of the standard unit, the value of **centi-** is $\frac{1}{100}$ of the standard unit, the value of **milli-** is $\frac{1}{1,000}$ of the standard unit, and so on.

• Decimal point

Thousands (1,000)	Hundreds (100)	Tens (10)	Units or ones (1)	Tenths ($\frac{1}{10}$)	Hundredths ($\frac{1}{100}$)	Thousandths ($\frac{1}{1,000}$)
Kilo-	Hecto-	Deka-	**STANDARD UNIT**	Deci-	Centi-	Milli-

There are other metric prefixes for very large and very small amounts. For measurements smaller than one-thousandth of a unit or larger than one thousand times a unit, prefixes that align with periods on a place-value chart are commonly used.

Metric Prefixes

Prefix	Meaning
atto-(a)	quintillionth part ($\times 10^{-18}$)
femto-(f)	quadrillionth part ($\times 10^{-15}$)
pico-(p)	trillionth part ($\times 10^{-12}$)
nano-(n)	billionth of ($\times 10^{-9}$)
micro-(μ)	millionth of ($\times 10^{-6}$)
milli-(m)	thousandth of ($\times 10^{-3}$)
centi-(c)	hundredth of ($\times 10^{-2}$)
deci-(d)	tenth of ($\times 10^{-1}$)
	Standard Unit
deka-/deca-(dk)	ten times ($\times 10^{1}$)
hecto-(h)	hundred times ($\times 10^{2}$)
kilo-(k)	thousand times ($\times 10^{3}$)
mega-(M)	million times ($\times 10^{6}$)
giga-(G)	billion times ($\times 10^{9}$)
tera-(T)	trillion times ($\times 10^{12}$)
peta-(P)	quadrillion times ($\times 10^{15}$)
exa-(E)	quintillion times ($\times 10^{18}$)

● **RELATED TOPICS: Place values, Powers of 10**

Length

Meter: The **meter** is the standard unit for measuring length. Both *meter* and *metre* are acceptable spellings for this unit of measure.

A meter is about 39.37 in., or 3.37 in. (approximately $3\frac{1}{3}$ in.) longer than a yard. We use the meter to measure lengths and distances like room dimensions, land dimensions, lengths of poles, heights of mountains, and heights of buildings. The abbreviation for meter is m.

Kilometer: A **kilometer** is 1,000 meters and is used for longer distances. The abbreviation for kilometer is km. The prefix *kilo* means 1,000. We measure the distance from one city to another, one country to another, or

Direct Measurement

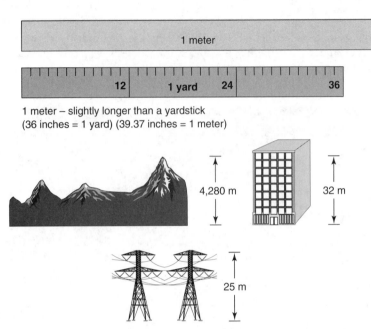

1 meter – slightly longer than a yardstick
(36 inches = 1 yard) (39.37 inches = 1 meter)

one landmark to another in kilometers. Driving at a speed of 55 mi (or 90 km) per hour, we would travel 1 km in about 40 sec. An average walking speed is 1 km in about 10 min.

1 kilometer — about five city blocks
1 mile — about eight city blocks

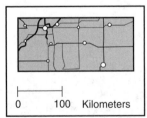

Centimeter: The **centimeter** is commonly used to measure objects less than 1 m long. The prefix *centi* means "$\frac{1}{100}$ of," and a centimeter is one hundredth of a meter. A centimeter is about the width of a thumbtack head, somewhat less than $\frac{1}{2}$ in. We use centimeters to measure medium-sized objects such as tires, clothing, textbooks, and television pictures.

1 cm

1 centimeter — about the width of a thumbtack or large paper clip

Millimeter: Many objects are too small to be measured in centimeters. A **millimeter** (mm) is used, which is "$\frac{1}{1,000}$ of" a meter. It is about the thickness of a plastic credit card or a dime. Certain film sizes, bolt and nut sizes, the length of insects, and similar items are measured in millimeters.

Other units and their abbreviations are decimeter, dm; dekameter, dkm; and hectometer, hm.

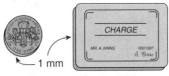

1 millimeter—about the thickness
of a dime or a charge card

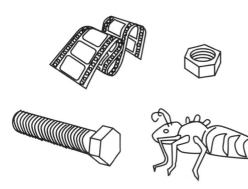

Weight or Mass

Gram: The standard unit for measuring mass in the metric system is the **gram.** A gram is the mass of 1 cubic centimeter (cm^3) of water at its maximum density. A cubic centimeter is a cube whose edges are each 1 centimeter long. It is a little smaller than a sugar cube. The abbreviation g is used for gram. Grams are used to measure small or light objects such as paper clips, food products, coins, and bars of soap. The metric unit for measuring weight is the Newton (N); however, in common usage the gram is used in comparing metric and U.S. customary units of weight.

1 gram — about the weight of two paper clips

Kilogram: A **kilogram** (kg) is 1,000 grams. Since a cube 10 cm on each edge can be divided into 1,000 cm^3, the weight of water required to fill this cube is 1,000 grams or 1 kilogram. A kilogram is approximately 2.2 lb. The kilogram is probably the most often used unit for weight. We use it to weigh people, meat, bags of flour, automobiles, and so on.

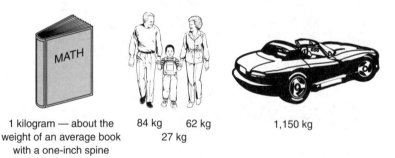

1 kilogram — about the   84 kg   62 kg   1,150 kg
weight of an average book   27 kg
with a one-inch spine

Milligram: The gram is used to measure small objects, and the **milligram** (1/1,000 of a gram) is used to measure *very* small objects. Milligrams are much too small for ordinary uses; however, pharmacists use milligrams (mg) to measure small amounts of drugs, vitamins, and medications.

1 milligram

Aspirin

Other units and their abbreviations are decigram, dg; centigram, cg; deka-gram, dkg; and hectogram, hg.

Capacity or Volume

Liter: A **liter** (L) is the volume of a cube 10 cm on each edge. It is the standard metric unit of capacity. Like the meter, it may be spelled *liter* or *litre*, but we use the spelling *liter*. A cube 10-cm on each edge filled with water weighs approximately 1 kg, so 1 L of water weighs about 1 kg. One liter is just a little larger than a liquid quart. Soft drinks are often sold in 2-liter bottles, gasoline is sold by the liter at some service stations, and numerous other products are sold in liter containers.

1 liter — about the volume of a quart of milk or a soft drink in a plastic bottle

A small car holds approximately 60 liters of gasoline.

Milliliter: A liter is 1,000 cm³, so $\frac{1}{1,000}$ of a liter, or a **milliliter**, has the same volume as a cubic centimeter. Most liquid medicine is labeled and sold in milliliters (mL) or cubic centimeters (cc or cm³). Medicines, perfumes, and other very small quantities are measured in milliliters.

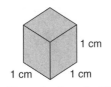

1 cubic centimeter = 1 milliliter

1 dose = 5 ml

Other units and their abbreviations are deciliter, dL; centiliter, cL; dekaliter, dkL; hectoliter, hL; and kiloliter, kL.

Other multiples and standard units in the metric system exist, but those discussed here are the ones frequently used.

How to

Change from one metric unit to any smaller metric unit:

1. Mentally position the measure on the metric-value chart so that the decimal immediately follows the original measuring unit.
2. Move the decimal to the right so that it *follows* the new measuring unit. (Attach zeros if necessary.)

EXAMPLE   To change 43 dkm to cm:
Place 43 dkm on the metric-value chart so that the last digit is in the dekameters place; that is, the understood decimal that follows the 3 will be *after* the dekameters place. To change to centimeters, move the decimal point so that it follows the centimeters place. Fill in the empty places with zeros.

Note the shortcut to multiply by powers of 10: Move the decimal point one place to the right for each time 10 is used as a factor.

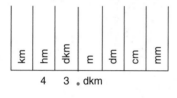

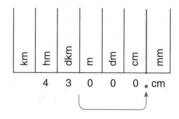

43 dkm = 43,000 cm.

How to

Change from one metric unit to any larger metric unit:

1. Mentally position the measure on the metric-value chart so that the decimal immediately follows the original unit.
2. Move the decimal to the left so that it *follows* the new unit. (Attach zeros if necessary.)

EXAMPLE   To change 3,495 L to kL:
Place the number on the chart so the digit 5 is in the liters place; that is, the decimal point follows the liters place. (Note the understood decimal point

after the liters place.) To change the kiloliters, move the decimal *three* places to the left so the decimal follows the kiloliters place. Use the shortcut to divide by powers of 10: Move the decimal point one place to the left for each time 10 is used as a divisor.

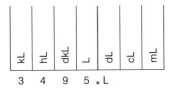

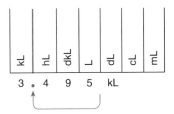

3,495 L = 3.495 kL.

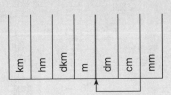

TIP

How far and which way?

To determine the movement of the decimal point when changing from one metric unit to another, answer these questions:

1. *How far* is it from the original unit to the new unit (how many places)?
2. *Which way* is the movement on the chart (left or right)?

Change 28.392 cm to m.

    How far is it from cm to m?   **Two places**
    Which way?   **Left**

Move the decimal in the original measure *two places to the left.*

    28.392 cm = 0.28392 m

Conversion factors can also be used to convert from one metric measure to another.

Metric System

	From	To	Multiply By
Length or Distance			
1 kilometer (km) = 1,000 meters (m)	kilometers	meters	1,000
	meters	kilometers	0.001
1 hectometer (hm) = 100 meters	hectometers	meters	100
	meters	hectometers	0.01
1 dekameter (dkm) = 10 meters	dekameters	meters	10
	meters	dekameters	0.1
1 decimeter (dm) = 0.1 meter	decimeters	meters	0.1
	meters	decimeters	10
1 centimeter (cm) = 0.01 meter	centimeters	meters	0.01
	meters	centimeters	100
1 millimeter (mm) = 0.001 meter	millimeters	meters	0.001
	meters	millimeters	1,000
Weight			
1 kilogram (kg) = 1,000 grams (g)	kilograms	grams	1,000
	grams	kilograms	0.001
1 hectogram (hg) = 100 grams	hectograms	grams	100
	grams	hectograms	0.01
1 dekagram (dkg) = 10 grams	dekagrams	grams	10
	grams	dekagrams	0.1
1 decigram (dg) = 0.1 gram	decigrams	grams	0.1
	grams	decigrams	10
1 centigram (cg) = 0.01 gram	centigrams	grams	0.01
	grams	centigrams	100
1 milligram (mg) = 0.001 gram	milligrams	grams	0.001
	grams	milligrams	1,000
Capacity			
1 kiloliter (kL) = 1,000 liters (l)	kiloliters	liters	1,000
	liters	kiloliters	0.001
1 hectoliter (hL) = 100 liters	hectoliters	liters	100
	liters	hectoliters	0.01
1 dekaliter (dkL) = 10 liters	dekaliters	liters	10
	liters	dekaliters	0.1
1 deciliter (dL) = 0.1 liter	deciliters	liters	0.1
	liters	deciliters	10
1 centiliter (cL) = 0.01 liter	centiliters	liters	0.01
	liters	centiliters	100
1 milliliter (mL) = 0.001 liter	milliliters	liters	0.001
	liters	milliliters	1,000

How to

Add or subtract like *or* common *metric measures:*

1. Add or subtract the numerical values.

   7 cm + 4 cm = 11 cm

2. Give the answer in the common unit of measure.

   15 dkg − 3 dkg = 12 dkg

How to

Add or subtract unlike *metric measures:*

1. Change the measures to a common unit of measure.
2. Add or subtract the numerical values.
3. Give the answer in the common unit of measure.

EXAMPLE   To add 9 mL + 2 cL:

   9 mL + 2 cL   Change cL to mL; that is, 2 cL = 20 mL.

   9 mL + 20 mL = **29 mL**

Alternate solution:

   9 mL + 2 cL   Change milliliters to centiliters. 9 mL = 0.9 cL

   0.9 cL + 2 cL =
   $$\begin{array}{r} 0.9 \text{ cL} \\ \underline{2 \quad \text{ cL}} \\ 2.9 \text{ cL} \end{array}$$ Note alignment of decimals.
                          An understood decimal follows the addend 2.

   Since 2.9 cL = 29 mL, the answers are equivalent.

EXAMPLE   To subtract 14 km − 34 hm:

   $$\begin{array}{r} 14.0 \text{ km} \\ -\quad 3.4 \text{ km} \\ \hline 10.6 \text{ km} \end{array}$$ Change hm to km; that is, 34 hm = 3.4 km.
                          Watch alignment of decimals.

   The difference is 10.6 km, or 106 hm.

**Direct
Measurement**

How to

Multiply a metric measure by a number:

1. Multiply the numerical values.
2. Give the answer in the same unit as the original measure.

$2.7 \text{ m} \times 3 = 8.1 \text{ m}$

How to

Divide a metric measure by a number:

1. Divide the numerical values.
2. Give the answer in the same unit as the original measure.

$30 \text{ cm} \div 5 = 6 \text{ cm}$

How to

Divide a metric measure by a like measure:

1. Divide the numerical values.
2. Give the answer as a number that tells how many parts there are.
3. If unlike measures are used, first change them to like measures.

$$\frac{250 \text{ mg}}{100 \text{ mg}} = 2.5$$

■ 18–4 Metric–U.S. Customary Comparisons

How to

Convert metric measures to U.S. customary measures or U.S. customary measures to metric measures:

1. Select the appropriate conversion factor.
2. Multiply and include the appropriate unit of measure.

U.S. Customary and Metric Comparisons

	From	To	Multiply By
Length			
1 meter = 39.37 inches	meters	inches	39.37
	inches	meters	0.0254
1 meter = 3.28 feet	meters	feet	3.2808
	feet	meters	0.3048
1 meter = 1.09 yards	meters	yards	1.0936
	yards	meters	0.9144
1 centimeter = 0.39 inch	centimeters	inches	0.3937
	inches	centimeters	2.54
1 millimeter = 0.04 inch	millimeters	inches	0.03937
	inches	millimeters	25.4
1 kilometer = 0.62 miles	kilometers	miles	0.6214
	miles	kilometers	1.6093
Weight			
1 gram = 0.04 ounce	grams	ounces	0.0353
	ounces	grams	28.3286
1 kilogram = 2.2 pounds	kilograms	pounds	2.2046
	pounds	kilograms	0.4536
Liquid Capacity			
1 liter = 1.06 quarts	liters	quarts	1.0567
	quarts	liters	0.9463

Direct Measurement

EXAMPLE   To change 50 ft to meters:

feet to meters: 0.3048   Conversion factor.

50 × 0.3048 = 15.24 m   Multiply.

└

▲ **APPLICATION: Body Mass Index**

Body mass index (BMI) is the standard unit for measuring a person's degree of obesity or emaciation. BMI is body weight in kilograms (kg) divided by height in meters squared, or BMI $= w/h^2$.

According to federal guidelines, a BMI greater than 25 means that you are overweight. Surveys show that 59% of men and 49% of women have BMIs greater than 25. Extreme obesity is defined as a BMI greater than 40.

How to

Calculate your body mass index (BMI):

1. Multiply your weight in pounds by 0.45 to convert to kilograms.
2. Convert your height to inches.
3. Multiply the inches by 0.025 to get meters.
4. Square this number.
5. Divide your weight in kilograms by this number, and round to the nearest whole number. The result is your BMI.

EXAMPLE   Alexa May is 5 feet 6 inches and weighs 138 pounds. To find her body mass index:

$138 \times 0.45 =$ 62.1   Change pounds to kilograms.

$5 \text{ ft } 6 \text{ in.} = 5 \times 12 + 6$ Change feet to inches.

$\qquad = 60 + 6$

$\qquad = 66 \text{ inches}$

$66 \times 0.025 =$ 1.65   Change inches to meters.

$\text{BMI} = \dfrac{w}{h^2}$ $w = 62.1, h = 1.65$

$\text{BMI} = \dfrac{62.1}{(1.65)^2}$

$\text{BMI} = \dfrac{62.1}{2.7225}$

$\text{BMI} = 22.8$

▉ 18–5 Reading Measuring Instruments

Many measuring instruments are available for making measurements of all types. On some measuring instruments we read the measurement directly from a scale, whereas on other measuring instruments that have electronic sensors we read the measurement on a digital display. Whether the measurement is made manually or electronically, all measurements are *approximate* values.

Approximate numbers that represent measured values may have varying degrees of accuracy. The **accuracy** of a measurement refers to how close the measured value is to the true or accepted value.

Precision is the degree to which several measurements provide values very close to one another.

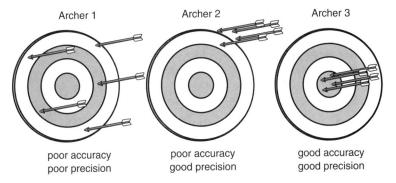

Archer 1
Archer 2
Archer 3

poor accuracy
poor precision

poor accuracy
good precision

good accuracy
good precision

One indicator of the degree of precision of a measurement is the number of **significant digits** in a measure. The numbers 2,500; 250; 25; 2.5; 0.25 and 0.025 all have two significant digits. When a number contains no zeros, all the digits are significant. However, zeros are special digits. Sometimes zeros are significant digits, and sometimes they are just placeholders.

Direct Measurement

How to

Determine the significant digits of a number:

For whole numbers:

Determine the number of significant digits in the following:

1. Start with the left-most nonzero digit.
2. Count each digit through the right-most nonzero digit.

200 has 1 significant digit.
28 has 2 significant digits.
1320 has 3 significant digits.
2005 has 4 significant digits.

For decimals and mixed decimals.

1. Start with the left-most nonzero digit.
2. Count each digit through the last digit.

0.07 has 1 significant digit.
2.0 has 2 significant digits.
1.20 has 3 significant digits.
0.01250 has 4 significant digits.

● RELATED TOPIC: Nonzero digit

How precise can we be?

The concept of precision is associated with approximate numbers. Exact numbers have infinite precision. The average measurement of 2.3 cm, 4.25 cm, and 5.125 cm can be no more precise than the least precise of the measurements (tenths).

For the calculations $\frac{2.3 + 4.25 + 5.125}{3}$, we would round the results to the nearest tenth. The number 3 is an exact number (exactly three measures) and has an infinite precision, so tenths is still the least precise.

$$\frac{2.3 + 4.25 + 5.125}{3} = \frac{11.675}{3} = 3.891666667 = 3.9 \text{ (rounded to tenths)}$$

The **greatest possible error** of a measurement is half of the precision of the measurement.

How to

Find the greatest possible error of a measurement:

$$1\frac{3}{4}$$

1. Determine the precision of the measurement of the smallest subdivision.

$$\text{precision} = \frac{1}{4}$$

2. Find half of the precision.

$$\frac{1}{2} \times \frac{1}{4} = \frac{1}{8}$$

EXAMPLE To find the greatest possible error of the measurements:

(a) $2\frac{5}{8}$ in.   (b) 3.5 cm

(a) $2\frac{5}{8}$ in.   Precision $= \frac{1}{8}$.

$$\frac{1}{2} \times \frac{1}{8} = \frac{1}{16} \text{ in.}$$ Greatest possible error.

(b) 3.5 cm   Precision $= 0.1$.

$$0.5(0.1) = 0.05 \text{ cm.}$$ Greatest possible error.

The **U.S. customary rule** uses an inch as the standard unit. Each inch is subdivided into fractional parts, usually 8, 16, 32, or 64.

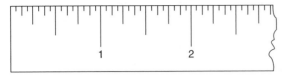

U.S. Customary Rule

In the illustration each inch is divided into 16 equal parts, so each part is $\frac{1}{16}$ in.; that is, the first mark from the left edge represents $\frac{1}{16}$ in. The left end of the rule represents zero (0). (Some rules leave a small space between zero and the end of the rule.)

The division marks are different lengths to make the rule easier to read. The shortest marks represent fractions that, in lowest terms, are sixteenths $\left(\frac{1}{16}, \frac{3}{16}, \frac{5}{16}, \frac{7}{16}, \frac{9}{16}, \frac{11}{16}, \frac{13}{16}, \frac{15}{16}\right)$. Thus, the marks representing fractions that reduce to eighths are slightly longer than the sixteenths marks $\left(\frac{1}{8}, \frac{3}{8}, \frac{5}{8}, \frac{7}{8}\right)$. Next, the marks representing fractions that reduce to fourths are slightly longer than the eighths $\left(\frac{1}{4}, \frac{3}{4}\right)$. The mark for one-half $\left(\frac{1}{2}\right)$ is longer than the fourths, and the inch marks are the longest.

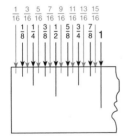

Direct Measurement

How to

Read a measure from a scaled rule:

1. Align one end of the line segment or item to be measured with zero on the rule.
2. Determine the mark that is closest to the other end of the line segment or item.
3. Determine the value represented by the mark identified in Step 2.

TIP

Judge to the closest mark

A line segment or item may not always align exactly with a division mark. If this is the case, use eye judgment to decide which mark is closer to the end of the line segment or item.

EXAMPLE   To measure line segment $\overline{AB}$:

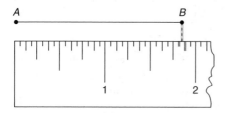

Point B aligns between $1\frac{13}{16}$ and $1\frac{7}{8}$. Recall that measurements are always approximations; using our best eye judgment, we determine that point B seems closer to $1\frac{13}{16}$ than to $1\frac{7}{8}$.

AB is $1\frac{13}{16}''$ to the nearest sixteenth of an inch.

Many standard rulers have both a U.S. customary scale and a metric scale. The metric rule is usually calibrated in centimeters or millimeters. The **metric rule** illustrated below shows centimeters (cm) as the major divisions, represented by the longest lines. Each centimeter is divided into 10 millimeters (mm), which are the shortest lines. A line slightly longer than the millimeter line divides each centimeter into two equal parts of 5 mm each.

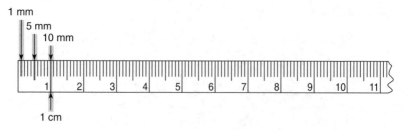

Metric Rule

EXAMPLE   To find the length of the line segment *CD*:

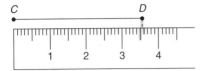

The end of line segment *CD* falls approximately halfway between the 35-mm mark and the 36-mm mark, by eye judgment, measuring about 35.5 mm. Both 35 and 36 mm would be acceptable measures of the line segment *CD*. **The measurements 35 mm, 35.5 mm, and 36 mm are acceptable approximations for line segment *CD*.**

◢▂

■ 18–6 Time Zones

Since 1 day is 24 hours, the Earth is divided into **time zones** based on a 24-hr day. To understand how time zones work, let's look at how locations are defined on a globe of the Earth. The grid system has similarities to the rectangular coordinate system, but since the Earth is approximately a sphere (a ball-like shape), there are also differences.

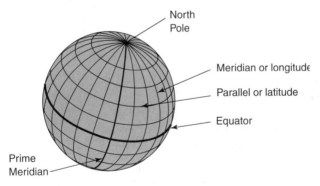

The imaginary horizontal line we call the **equator** divides the Earth into two halves called **hemispheres**. Additional imaginary horizontal circles or rings are called **parallels of latitude**. Since the equator is our north/south reference, we define it as 0° on the scale. The parallels number from 0° to 90° both north and south. The North Pole is at 90° N, and the South Pole is at 90° S. We can relate this scale to the rectangular coordinate system and the vertical axis. Northern latitudes correspond to positive y-values on the rectangular coordinate system. Southern latitudes correspond to negative y-values.

Imaginary circles drawn around the globe that pass through both the North and South Poles are called **meridians of longitude.** Unlike the parallels, meridians are farthest apart at the equator and closest together at the North and South Poles. They roughly define time zones.

One complete rotation is 360°, so one time zone covers approximately 15° of the Earth's rotation (360° ÷ 24 = 15°). We say *approximately* because time zone boundaries are not defined strictly by meridians. Actual time zone boundaries are defined by country, state, or province boundaries that are near a particular meridian. A few counties or small islands that are close to halfway between time zone boundaries define their time as a $\frac{1}{2}$-hour difference from a neighboring time zone. To be sure how a country defines its time, consult an atlas or other reference. Many software programs, such as Microsoft Outlook℠, give this information. Just as the equator is the longest parallel and is the zero line that separates north and south measures, one meridian must serve as zero for the east and west measures. All meridians are the same length, so the one that passes through Greenwich, England, which is called the **prime meridian** was arbitrarily selected.

As the prime meridian passes through the North and South Poles and becomes the 180° meridian on the side of the Earth opposite Greenwich, England, it is called the **international date line.** In references that show the appropriate time zone for a location (city, state, or country), the time is given as plus or minus a specified number of hours from **Greenwich Mean Time (GMT).**

▲ APPLICATION: Time Zones

How to

Determine the time in a different time zone:

1. Identify the GMT for both locations.
2. Subtract the GMT for the location with the known time from the GMT for the location with the unknown time.
3. Adjust the known time by the number of hours found in Step 2.

EXAMPLE   To determine the time in Anchorage, Alaska, USA when it is 3:00 P.M. in Caracas, Venezuela:

Anchorage: −10 GMT   Identify each GMT.

Caracas: −4 GMT

$-10 - (-4)$ Subtract GMT for unknown $-$ known.

$= -10 + 4 = -6$ Interpret as 6 hours earlier.

3:00 P.M. $= 15:00$ Time on a 24-hr clock.

$15:00 - 6:00 =$ **9:00** A.M.

∟

Subtracting a negative

Dealing with time zones is a practical application for signed number skills. When we calculated the difference between Caracas time and Anchorage time, we got $-10 - (-4) = -10 + 4 = -6$.

Our answer of -6 means that it is 6 hours *earlier* in Anchorage than it is in Caracas. If we instead find the difference between Anchorage time and Caracas time $(-4 - (-10) = -4 + 10 = +6)$, our answer of $+6$ means that it is 6 hours *later* in Caracas than it is in Anchorage.

To minimize errors, anticipate whether the time should be earlier or later in the second location.

● **RELATED TOPICS: Adding integers, Subtracting integers**

Direct
Measurement

Time Zone Chart

GMT	Military	Phonetic	Civilian Time Zones	Cities/Country
+0:00	z	Zulu	GMT—Greenwich Mean UT or UTC—Universal (Coordinated) WET—Western European	London, England Dublin, Ireland Edinburgh, Scotland Lisbon, Portugal Reykjavik, Iceland Casablanca, Morocco
−1:00	a	Alpha	WAT—West Africa	Azores, Cape Verde Islands
−2:00	b	Bravo	AT—Azores	Brasilia, Brazil
−3:00	c	Charlie		Buenos Aires, Argentina Georgetown, Guyana
−4:00	d	Delta	AST—Atlantic Standard	Caracas, Venezuela La Paz, Bolivia
−5:00	e	Echo	EST—Eastern Standard	Atlanta, GA, USA Bogota, Colombia Boston, MA, USA Lima, Peru New York, NY, USA Toronto, Canada

Offset	Letter	Phonetic	Zone	Locations
−6:00	f	Foxtrot	CST—Central Standard	Chicago, IL, USA; Dallas, TX, USA; Mexico City, Mexico; New Orleans, LA, USA; Saskatchewan, Canada
−7:00	g	Golf	MST—Mountain Standard	Denver, CO, USA
−8:00	h	Hotel	PST—Pacific Standard	Los Angeles, CA, USA
−9:00	j	Juliet	YST—Yukon Standard	Whitehorse, Yukon, Canada; Juneau, Alaska, USA
−10:00	k	Kilo	AHST—Alaska–Hawaii Standard; CAT—Central Alaska; HST—Hawaii Standard; EAST—East Australian Standard	Honolulu, Hawaii, USA; Anchorage, Alaska, USA; Fairbanks, Alaska, USA
−11:00	l	Lima	NT—Nome	Nome, Alaska, USA; American Samoa; Aleutian Islands
−12:00	m	Mike	IDLW—International Date Line West	
+12:00	y	Yankee	IDLE—International Date Line East; NZST—New Zealand Standard; NZT—New Zealand; Russia Zone 10	Attu, Alaska, USA; Wellington, New Zealand; Fiji; Marshall Islands; Tasmania;
+11:00	x	X-ray		Sydney, Australia; Micronesia; Tamuning, Guam;
+10:00	w	Whiskey	GST—Guam Standard, Russia Zone 9	Vladivostok; Papua, New Guinea

Direct Measurement

Time Zone Chart (continued)

GMT	Military	Phonetic	Civilian Time Zones	Cities/Country
+9:00	v	Victor	JST—Japan Standard, Russia Zone 8	Korea, Singapore, Saipan, Tokyo, Malaysia
+8:00	u	Uniform	CCT—China Coast, Russia Zone 7	Mongolia; Perth, Australia
+7:00	t	Tango	WAST—West Australian Standard, Russia Zone 6	Christmas Island, Australia; Java, Indonesia, Vietnam, Laos
+6:00	s	Sierra	ZP6—Chesapeake Bay, Russia Zone 5	Omsk, Kazakhstan (East), Kyrgyzstan
+5:00	r	Romeo	ZP5—Chesapeake Bay, Russia Zone 4	Perm, Pakistan; Kazakhstan (West)
+4:00	q	Quebec	ZP4—Russian Zone 3	Mauritius/Abu Dhabi, UAE Muscat, Aman Tblisi, Republic of Georgia Volgograd, Russia Kabul, Afghanistan
+3:00	p	Papa	BT—Baghdad, Russia Zone 2	Kuwait Nairobi, Kenya Riyadh, Saudi Arabia Moscow, Russia Tehran, Iran

+2:00	o	Oscar	EET—Eastern European, Russia Zone 1	Athens, Greece Helsinki, Finland Istanbul, Turkey Jerusalem, Israel Botswana, Zambia, Mozambique Harare, Zimbabwe
+1:00	n	November	CET—Central European FWT—French Winter MET—Middle European MEWT—Middle European Winter SWT—Swedish Winter	Paris, France Berlin, Germany Amsterdam, The Netherlands Brussels, Belgium Vienna, Austria Madrid, Spain Rome, Italy Bern, Switzerland Stockholm, Sweden Oslo, Norway

Direct
Measurement

CHAPTER 19

Perimeter, Area, and Volume

- ■ **19–1 Perimeter and Area**
- ■ **19–2 Circles and Composite Shapes**
- ■ **19–3 Three-Dimensional Figures**

■ 19–1 Perimeter and Area

A **polygon** is a plane or flat, closed figure described by straight-line segments and angles. Polygons have different numbers of sides and different properties. Some common polygons are the parallelogram, rectangle, square, trapezoid, and triangle.

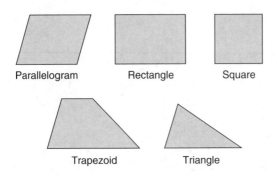

Parallelogram   Rectangle   Square

Trapezoid   Triangle

    A **parallelogram** is a four-sided polygon with opposite sides that are parallel.

    A **square** is a parallelogram with sides all of equal length and with all right angles.

    A **rectangle** is a parallelogram with angles that are all right angles.

A **trapezoid** is a four-sided polygon having only two parallel sides.
A **triangle** is a polygon that has three sides.
The **perimeter** is the total length of the sides of a plane figure.

● **RELATED TOPICS: Line, Plane, Point**

When do you need to find the perimeter of a polygon? Some common applications for perimeter are finding the amount of trim molding for a room, determining the amount of fencing for a yard, finding the amount of edging for a flower bed, and so on.

A general procedure for finding the perimeter of any shape is to add the lengths of the sides. However, shortcuts based on the properties of the shape are given as formulas.

	Perimeter		
Square	$P = 4s$		s is length of a side
Rectangle	$P = 2l + 2w$ or $P = 2(l + w)$		l is length w is width
Parallelogram	$P = 2b + 2s$ or $P = 2(b + s)$		b is the base s is the adjacent side
Trapezoid	$P = a + b + c + d$		$a, b, c,$ and d are sides
Triangle	$P = a + b + c$		$a, b,$ and c are sides

How to

Find the perimeter of a polygon:

1. Select the appropriate formula.
2. Substitute the known values into the formula.
3. Evaluate the formula.

EXAMPLE   To find the perimeter of a parallelogram with a base of 16 in. and an adjacent side of 8 in.:
Visualize the parallelogram.

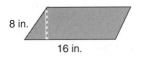

8 in.

16 in.

$P_{parallelogram} = 2(b + s)$ Select the perimeter of parallelogram formula.

$P_{parallelogram} = 2(16 + 8)$ Substitute: $b = 16, s = 8$.

$P_{parallelogram} = 2(24)$ Evaluate.

$P_{parallelogram} = 48$ in.

The **area of a polygon** is the amount of surface of a plane figure. Area is expressed in square units. For example, if a rectangle is 3 meters wide and 5 meters long, there are 15 square meters in the area.

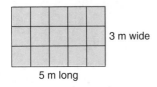

3 m wide

5 m long

Area $= 15$ m^2

Writing square units

The exponent 2 following a unit of measure indicates *square measure* or area. Thus, 15 m^2 = 15 square meters, 23 ft^2 = 23 square feet, 120 cm^2 = 120 square centimeters, and so on. This is a shortcut way to express a measure that has *already* been "squared."

When do you need the area of a polygon? Some common applications for area are finding the amount of carpeting needed, the amount of paint needed to paint a surface, the amount of fertilizer needed to treat a lawn, and so on.

Area			
Square	$A = s^2$		s is length of a side
Rectangle	$A = lw$		l is length w is width
Parallelogram	$A = bh$		b is base h is height
Trapezoid	$A = \frac{1}{2}h(b_1 + b_2)$		b_1 and b_2 are bases h is height
Triangle	$A = \frac{1}{2}bh$		b is base h is height

How to

Find the area of a polygon:

1. Visualize the polygon.
2. Select the appropriate formula.
3. Substitute the known values into the formula.
4. Evaluate the formula.

▲ **APPLICATION: Area of a House Gable**

EXAMPLE   A gable on a house has a rise of 7 ft 6 in. and a span of 30 ft 6 in. The rise is the height of the gable, and the span is the base of the gable. To find the area:

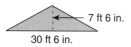
— 7 ft 6 in.
30 ft 6 in.

Geometry

$A = \dfrac{1}{2}bh$ Substitute in formula after converting to one common unit of measure. 7 ft 6 in. = 7.5 ft, 30 ft 6 in. = 30.5 ft because 1 ft = 12 in.

$A = \dfrac{1}{2}(30.5)(7.5)$ Change $\frac{1}{2}$ to 0.5.

$A = \boxed{0.5}\,(30.5)(7.5)$ Multiply.

$A = 114.375 \text{ ft}^2$

The area of the gable is 114.375 ft².

When the three sides of a triangle are known and the height is not known, the area of a triangle can be found using a different formula. The formula is known as *Heron's formula.*

How to

Find the area of a triangle when the length of the three sides are known:

1. Use the formula:

    $\text{Area} = \sqrt{s(s-a)(s-b)(s-c)}$ Heron's formula

    where $s = \frac{1}{2}(a+b+c)$ and a, b, and c are the lengths of the three sides.
2. Find s.
3. Substitute values for a, b, c, and s into the formula.
4. Evaluate the formula.

EXAMPLE   To find the area of the triangle by using Heron's formula:

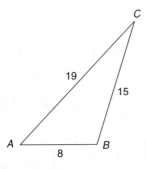

$s = \dfrac{1}{2}(a + b + c)$ Find s. Substitute length of sides.

$s = \dfrac{1}{2}(19 + 15 + 8)$ Add, then multiply.

$s = 21$

Area $= \sqrt{s(s - a)(s - b)(s - c)}$ Heron's formula.

Area $= \sqrt{21(21 - 19)(21 - 15)(21 - 8)}$ Substitute for s, a, b, and c, and perform each subtraction.

Area $= \sqrt{21(2)(6)(13)}$ Multiply.

Area $= \sqrt{3{,}276}$ Take the square root.

Area $= 57.23635209$ Principal square root.

Area $= 57.24$ square units   Rounded to four significant digits.

● **RELATED TOPICS: Order of operations, Square root**

■ 19–2 Circles and Composite Shapes

A **circle** is a closed curved line with points that lie in a plane and are the same distance from the *center* of the figure.

The **center** of a circle is the point that is the same distance from every point on the circumference of the circle.

The **radius** (plural: *radii*, pronounced "ray · dē · ī") is a straight line segment from the center of a circle to a point on the circle. It is half the diameter.

The **diameter** of a circle is a straight line segment from a point on the circle through the center to another point on the circle.

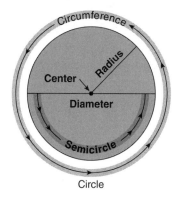

Circle

Geometry

The **circumference** of a circle is the perimeter or length of the closed curved line that forms the circle.

A **semicircle** is half a circle and is created by drawing a diameter.

The circle is a geometric form with a special relationship between its circumference and its diameter. If we divide the circumference of any circle by its diameter, the quotient is always the same number.

$$\pi = \frac{\text{Circumference}}{\text{Diameter}} = \frac{C}{d}$$

This number is a nonrepeating, nonterminating decimal approximately equal to 3.1415927 to seven decimal places. The Greek letter π (pronounced "pie") represents this value. Convenient approximations often used in calculations involving π are $3\frac{1}{7}$ and 3.14. Many calculators have a π key.

The formulas for the circumference and area of a circle are:

Circumference (C)	Area (A)		
$C = \pi d$ or $C = 2\pi r$	$A = \pi r^2$	d is diameter ($d = 2r$) r is radius ($r = \frac{1}{2}d$)	

How to

Find the circumference or area of a circle:

1. Select the appropriate formula.
2. Substitute values for r or d as appropriate.
3. Evaluate the formula. Use 3.14 or the calculator value for π.

EXAMPLE   To find the circumference of a circle that has a diameter of 1.3 m:

1.3 m

$C = \pi d$ Select circumference formula with diameter d.

$C = 3.14(1.3)$ Use 3.14 for π.

$C = 4.082$ m   Evaluate.

The circumference is 4.1 m (rounded).   Circumference is a linear measure.

EXAMPLE   To find the area of a circle that has a radius of 8.5 m:

$A = \pi r^2 = \pi(8.5)^2$ Select formula and substitute values.

$A = 3.14(72.25) = 226.865$ Evaluate using 3.14 for π.

8.5 m

The area of the circle is 227 m² (rounded).

A **composite figure** is a geometric figure made up of two or more geo-metric figures.

How to

Find a missing dimension of a composite shape:

1. Determine how the missing dimension is related to known dimensions.
2. Make a calculation of known dimensions according to the relationship found in Step 1.

EXAMPLE   To find the missing dimensions x and y on the slab foundation:

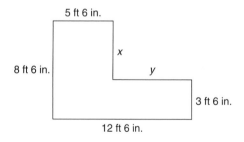

5 ft 6 in.

x

8 ft 6 in. y

3 ft 6 in.

12 ft 6 in.

Separate the figure into parts.

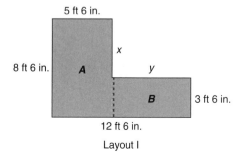

5 ft 6 in.

x

8 ft 6 in.   **A** y

B   3 ft 6 in.

12 ft 6 in.

Layout I

Geometry

In layout I, the side of B opposite its $3'6''$ side is also $3'6''$ because opposite sides of a rectangle are equal. The side of A opposite its $8'6''$ side is, for the same reason, $8'6''$. Dimension x must therefore be the difference between $8'6''$ and $3'6''$.

$x = 8'6'' - 3'6''$

$x = 5'$

5 ft 6 in.

8 ft 6 in.

C x

y

D   3 ft 6 in.

12 ft 6 in.

Layout II

If we think of layout II as two horizontal rectangles, we can find dimension y. The side opposite the $5'6''$ side of C must be $5'6''$. The side of D opposite the $12'6''$ side must also be $12'6''$. Dimension y must therefore be the difference between $12'6''$ and $5'6''$.

$y = 12'6'' - 5'6''$

$y = 7'$

The missing dimensions are $x = 5'$ and $y = 7'$.

⌐

▲ APPLICATION: Estimating Carpeting

Carpeting is most often purchased in square yards. However, measurements for a room are often given in feet.

How to

Estimate the amount of carpeting needed to cover a floor:

1. Sketch the area to be carpeted with the appropriate dimensions in feet.
2. Calculate the area.
3. Convert square feet to square yards.

EXAMPLE   To find the number of square yards of carpeting required for the room:

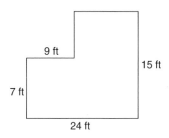

1. Divide the composite shape into two polygons with areas that can be computed. In this case, A is a rectangle and B is a square. Fill in the missing dimensions.

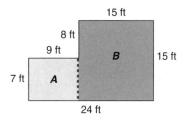

2. We find the areas of the smaller polygons and add them.

   Rectangle **A**   Square **B**

   $A_1 = lw$ $A_2 = s^2$

   $A_1 = 9 \times 7$ $A_2 = 15^2$

   $A_1 = \boxed{63 \text{ ft}^2}$ $A_2 = \boxed{225 \text{ ft}^2}$

   $\boxed{A_1} + \boxed{A_2} = \text{total area}$

   $\boxed{63} + \boxed{225} = 288 \text{ ft}^2$

3. We convert square feet to square yards using a unity ratio.

$$\frac{\overset{32}{\cancel{288}} \text{ ft}^2}{1} \times \frac{1 \text{ yd}^2}{\underset{1}{\cancel{9}} \text{ ft}^2} = 32 \text{ yd}^2$$

   The room requires **32 yd² of carpeting.**

▲ APPLICATION: Wallpapering a Room

Most wallpaper comes in 27-in.-wide rolls with 35 ft^2 per single roll. Paperers cut panels equal to the height of the room and cut out doors and windows from each panel. To find the number of single rolls needed, professional wallpaper hangers typically divide the total area to be covered by 22 ft^2. This technique ensures the minimum number of seams after allowing for door and window waste and pattern matching. How many single rolls of wallpaper will be needed using this method of calculation for the room in the drawings?

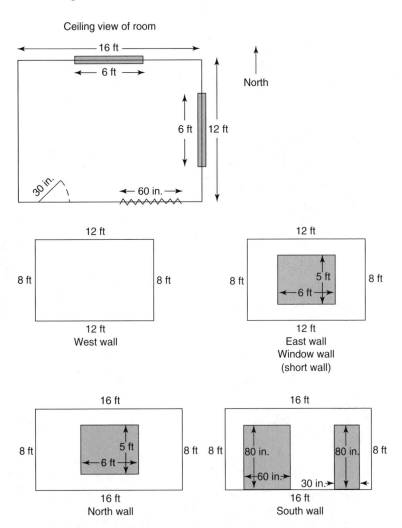

West wall

$A = lw$ $l = 12$ ft, $w = 8$ ft

$A = 12(8)$

$A =$ 96 ft^2

East wall

$A = lw$

Area of wall to be papered = Area of wall − Area of window

Area of wall = $12(8) = 96$ ft^2

Area of window = $6(5) = 30$ ft^2

Area to be papered = $96 − 30$

Area to be papered = 66 ft^2

North wall

Area = lw

Area of wall to be papered = Area of wall − Area of window

Area of wall = $16(8) = 128$ ft^2

Area of window = $6(5) = 30$ ft^2

Area of wall to be papered = $128 − 30 =$ 98 ft^2

East wall

Area = lw

Area of wall to be papered = Area of wall − $\begin{array}{c}\text{Area of} \\ \text{small door}\end{array}$ − $\begin{array}{c}\text{Area of} \\ \text{large door}\end{array}$

Area of wall = $16(8) = 128$ ft^2

Area of small door = $\dfrac{80}{12}\left(\dfrac{30}{12}\right) = \dfrac{2400}{144} = 16.67$ ft^2

Area of large door = $\dfrac{80}{12}\left(\dfrac{60}{12}\right) = \dfrac{4800}{144} = 33.33$ ft^2

Area to be papered = $128 − 16.67 − 33.33 =$ 78 ft^2

Geometry

Area of all walls
in the room to   =   Area of   +   Area of   +   Area of   +   Area of
be papered   west wall   east wall   north wall   south wall

Area = 96 + 66 + 98 + 78

Area of walls to be covered = 338 ft²

Number of rolls of wallpaper needed = $\dfrac{\text{Number of square feet (area) to be covered}}{22 \text{ ft}^2}$

Number of rolls = $\dfrac{338 \text{ ft}^2}{22 \text{ ft}^2}$

Number of rolls = 15.36 or 16

The room requires 16 rolls of wallpaper.

■ 19–3 Three-Dimensional Figures

Common household items like ice cubes, cardboard storage boxes, and toy building blocks are examples of three-dimensional geometric figures classified generally as *prisms*. Cans and pipes are examples of *cylinders*.

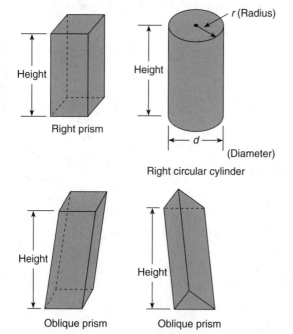

Right prism

Right circular cylinder

Oblique prism   Oblique prism

A **prism** is a three-dimensional figure with bases (ends) that are parallel, congruent polygons and faces (sides) that are parallelograms, rectangles, or squares. In a **right prism,** the faces are perpendicular to the bases.

A right circular **cylinder** is a three-dimensional figure with a curved surface and two circular bases such that the height is perpendicular to the bases.

In an **oblique prism** or an **oblique cylinder** the faces are *not* perpendicular to the bases.

The **height** of a three-dimensional figure with two bases is the shortest distance between the two bases.

In right circular cylinders and in right prisms, the height is the same as the length of a side or face. However, in oblique prisms and cylinders the height is the perpendicular distance between the bases and is different from the length of a side or face.

Other three-dimensional figures that are often used are the sphere and cone.

A **sphere** is a geometric figure formed by a curved surface with points that are all equidistant from a point inside called the **center.**

The **great circle** divides the sphere in half and is formed by a plane through the center of the sphere.

A **hemisphere** is one half of a sphere.

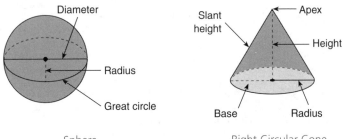

Sphere   Right Circular Cone

A **right cone** is a geometric figure that has a circular base. Its side surface tapers to a point, called the **vertex** or **apex,** and its height is a perpendicular line segment between the base and apex.

The **slant height** of a cone is the distance along the side from the base to the apex.

The **lateral surface area** (LSA) of a three-dimensional figure is the area of its sides only.

The **total surface area** (TSA) of a three-dimensional figure is the area of the sides plus the area of its base or bases.

The **volume** of a three-dimensional geometric figure is the amount of space it occupies, measured in terms of cubic measures.

The formulas for the lateral surface area, total surface area, and volume are:

	Lateral Surface Area	Total Surface Area	Volume
Right Rectangular Prism	$LSA = (2l + 2w)h$	$TSA = (2l + 2w)(h) + 2lw$	$V = lwh$
Right Circular Cylinder	$LSA = (\pi d)h$	$TSA = (\pi d)(h) + 2\pi r^2$	$V = \pi r^2 h$
Sphere	*	$TSA = 4\pi r^2$	$V = \dfrac{4\pi r^3}{3}$
Right Circular Cone	$LSA = \pi rs$	$TSA = \pi rs + \pi r^2$	$V = \dfrac{\pi r^2 h}{3}$

*A sphere does not have bases like prisms and cylinders. The surface area of a sphere includes *all* the surface so there is only one formula. Because of the relationship of the sphere to the circle, the formula includes elements of the formula for the area of a circle. The total surface area of the sphere is 4 times the area of a circle with the same radius.

▲ APPLICATION: Surface Area and Volume of Three-Dimensional Figures

How to

Find the surface area or volume of a three-dimensional figure:

1. Select the appropriate formula.
2. Substitute known values.
3. Evaluate the formula.

EXAMPLE   To find the lateral surface area, total surface area, and volume of a right circular cone with a diameter of 8 cm, height of 6 cm, and slant height of 7 cm:

$\text{LSA} = \pi rs$	Formula for LSA of a cone. Substitute values: $r = \frac{1}{2}d$, or 4.
$\text{LSA} = (3.14)(4)(7)$	Evaluate.
$\textbf{LSA} = \textbf{87.92 cm}^2$	
$\text{TSA} = \pi rs + \pi r^2$	Formula for TSA of a cone. Substitute values.
$\text{TSA} = 87.92 + (3.14)(4)^2$	Evaluate.
$\textbf{TSA} = \textbf{138.16 cm}^2$	Rounded.
$V = \dfrac{\pi r^2 h}{3}$	Formula for volume of a cone. Substitute values.
$V = \dfrac{(3.14)(4)^2(6)}{3}$	Evaluate.
$\textbf{V} = \textbf{100.48 cm}^3$	Rounded.

⌐

● **RELATED TOPIC: Order of operations**

▲ **APPLICATION: Ordering Concrete**

Concrete is purchased in cubic yards. Most often the project requiring the concrete is measured in inches and feet. The inches and feet can be converted to yards before calculating cubic yards, or the measurements can be converted to feet or inches and then from cubic feet or cubic inches to cubic yards. Fractional parts of a cubic yard are rounded up.

EXAMPLE   To find the amount of concrete required for a driveway that is 40 feet long and 9 feet wide if the concrete is to be 4 inches thick:

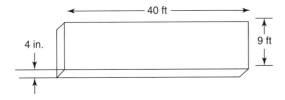

Geometry

Change all units to yards

$$40 \text{ ft}\left(\frac{1 \text{ yd}}{3 \text{ ft}}\right) = \frac{40}{3} \text{ yd} = 13.3333 \text{ yd}$$

$$9 \text{ ft}\left(\frac{1 \text{yd}}{3 \text{ ft}}\right) = 3 \text{ yd}$$

$$4 \text{ in.}\left(\frac{1 \text{ yd}}{36 \text{ in.}}\right) = \frac{4}{36} \text{ yd} = 0.1111 \text{ yd}$$

$$\text{Volume} = lwh$$

$$V = (13.3333)(3)(0.1111)$$

$$V = 4.4 \text{ yd}^3 \text{ or } \mathbf{5 \text{ yd}^3}$$

The driveway requires 5 cubic yards of concrete.

Alternative method:

Change all units to inches

$$40 \text{ ft}\left(\frac{12 \text{ in.}}{1 \text{ ft}}\right) = 480 \text{ in.} \qquad 9 \text{ ft}\left(\frac{12 \text{ in.}}{1 \text{ ft}}\right) = 108 \text{ in.}$$

$$V = lwh$$

$$V = 480(108)(4)$$

$$V = 207,360 \text{ in}^3$$

Change cubic inches to cubic yards

$$1 \text{ cubic yard} = (36 \text{ in.})^3 = 46,656 \text{ in}^3$$

$$V = 207,360 \text{ in}^3 \left(\frac{1 \text{ yd}^3}{46,656 \text{ in}^3}\right) = 4.44 \text{ yd}^3 \text{ or } \mathbf{5 \text{ yd}^3}$$

CHAPTER 20

Lines, Angles, and Polygons

- **20–1 Lines and Angles**
- **20–2 Triangles**
- **20–3 Inscribed and Circumscribed Regular Polygons and Circles**

■ 20–1 Lines and Angles

Geometry is the study of size, shape, position, and other properties of the objects around us. The basic terms used in geometry are *point, line,* and *plane.*

A **point** is a location or position that has no size or dimension. A dot is used to represent a point, and a capital letter is usually used to label the point.

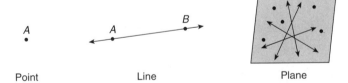

Point   Line   Plane

A **line** extends indefinitely in both directions and contains an infinite number of points. It has length but no width. The word *line* refers to a straight line unless otherwise specified. The line can be identified by naming any two points on the line (such as *A* and *B*).

A **plane** is a flat, smooth surface that extends indefinitely in all directions. A plane contains an infinite number of points and lines.

A **line segment,** or **segment,** consists of all points on the line between and including two points that are called **end points.**

Line segment $\overline{AB}$.

The notation for a line that extends through points A and B is $\overleftrightarrow{AB}$ (read "line AB"). The notation for the line segment including points A and B and all the points between is $\overline{AB}$ (read "line segment AB").

Another term used in connection with parts of a line is *ray*. Consider the beam of light from a flashlight. The beam is like a ray. It continues indefinitely in only one direction.

A **ray** consists of a point on a line and all points of the line on one side of the point.

The point from which the ray originates is called the **end point,** and all other points on the ray are called **interior points** of the ray. A ray is named by its end point and any interior point on the ray. The notation $\overrightarrow{RS}$ is used to denote the ray whose end point is R and that passes through S.

A line can be extended indefinitely in either direction. If two lines are drawn in the same plane, three things can happen:

1. The two lines **intersect** in *one and only one point.* $\overleftrightarrow{AB}$ and $\overleftrightarrow{CD}$ intersect at point E.

2. The two lines **coincide;** that is, one line fits exactly on the other. $\overleftrightarrow{EF}$ and $\overleftrightarrow{GH}$ coincide.

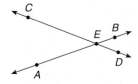

3. The two lines never intersect. In the following figure, $\overleftrightarrow{IJ}$ and $\overleftrightarrow{KL}$ are the same distance from each other along their entire lengths and are called **parallel lines.** The symbol $\parallel$ is used for parallel lines.

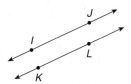

When two lines intersect in a point, four *angles* are formed. An **angle** is a geometric figure formed by two rays that intersect in a point, and the point of intersection is the end point of each ray.

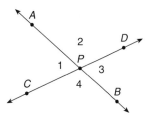

Rays $\overrightarrow{AB}$ and $\overrightarrow{AC}$ intersect at point A. Point A is the end point of $\overrightarrow{AB}$ and $\overrightarrow{AC}$. $\overrightarrow{AB}$ and $\overrightarrow{AC}$ are called the **sides** or **legs** of the angle. Point A is called the **vertex** of the angle.

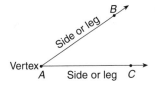

Angles can be named using a number or lowercase letter, by the capital letter that names the vertex point, or by using three capital letters. If three capital letters are used, two of the letters name interior points of each of the two rays, and the middle letter names the vertex point of the angle. Using the symbol $\angle$ for angle, the angle in the following figure can be named $\angle 1$, $\angle KLM$, $\angle MLK$, or $\angle L$. If one capital letter is used to name an angle, the letter is always the vertex letter. If three letters are used, the vertex letter is the center letter. The angle on the right is named with three letters as $\angle XZY$ or $\angle YZX$.

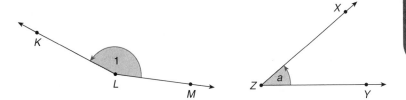

The measure of an angle is determined by the amount of opening between the two sides of the angle. Two units commonly used to measure angles are *degrees* and *radians*.

Consider the hands of a clock as the sides of an angle. When the two hands both point to the same number, the measure of the angle formed is 0 degrees (0°). An angle of 0 degrees is seldom used in geometric applications. During 1 hour, the minute hand makes one complete revolution. If we ignore the movement of the hour hand, this revolution of the minute hand contains 360 degrees (360°).

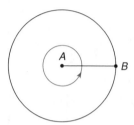

Rotation of 360°

A is kept as a fixed point and *B* rotates around point *A* then back to its original position. This rotation can be either clockwise or counterclockwise. A *complete rotation counterclockwise* is +360°. A *complete rotation clockwise* is −360°.

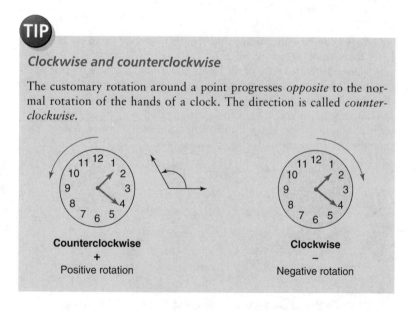

TIP

Clockwise and counterclockwise

The customary rotation around a point progresses *opposite* to the normal rotation of the hands of a clock. The direction is called *counterclockwise*.

Counterclockwise
+
Positive rotation

Clockwise
−
Negative rotation

A **degree** is one unit for measuring angles. It represents $\frac{1}{360}$ of a complete rotation about the vertex.

Suppose that $\overrightarrow{AC}$ rotates from $\overrightarrow{AB}$ through one-fourth of a circle. Then $\overrightarrow{AC}$ and $\overrightarrow{AB}$ form a 90° angle ($\frac{1}{4}$ of 360 = 90). This angle is a **right angle.** The symbol for a right angle is ∟ .

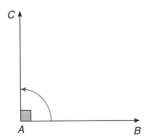

If two lines intersect so that right angles are formed (90° angles), the lines are **perpendicular** to each other. The symbol for "perpendicular" is ⊥.

If a string is suspended at one end and weighted at the other, it forms a **vertical line.** A line that is perpendicular to a vertical line is a **horizontal line.** $\overleftrightarrow{AB}$ is a vertical line, and $\overleftrightarrow{AB}$ and $\overleftrightarrow{CD}$ form right angles. $\overleftrightarrow{CD}$ is a horizontal line.

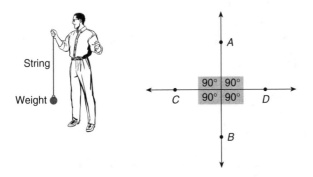

A **right angle** (90°) represents one-fourth of a circle or one-fourth of a complete rotation.

A **straight angle** (180°) represents half of a circle or half of a complete rotation.

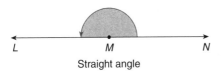

Straight angle

An **acute angle** is less than 90° but more than 0°.

Acute angle

An **obtuse angle** is more than 90° but less than 180°.

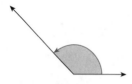

Obtuse angle

How to

Classify angles according to size:

1. Examine the angle to determine the number of degrees it measures.
2. Classify the angle as a straight, right, obtuse, or acute angle based on its degree measure.

EXAMPLE To classify the angles according to size:

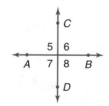

∠1 is acute.

∠2 is obtuse.

∠3 is acute.

∠4 is obtuse.

$\overleftrightarrow{AB} \perp \overleftrightarrow{CD}$

∠5, ∠6, ∠7, and ∠8

are all right angles.

∠ 9 is a straight angle.

Complementary angles are two angles that have a sum equivalent to one right angle or 90°.

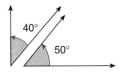

Complementary angles

Supplementary angles are two angles that have a sum equivalent to one straight angle or 180°.

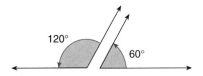

Supplementary angles

How to

Find the complement or supplement of an angle:

1. To find the complement of an angle, subtract its measure from 90°.
2. To find the supplement of an angle, subtract its measure from 180°.

EXAMPLE   To find the complement and supplement of an angle that measures 57°:

Complement   Supplement

90° − 57° = 33°   180° − 57° = 123°

The complement of 57° is 33°.   **The supplement of 57° is 123°.**

A **degree** is divided into 60 equal parts. Each part is called a **minute**.

1 degree (1°) = 60 minutes (60′)

The symbol ′ is used for minutes. Similarly, one minute is divided into 60 equal parts called **seconds.**

1 minute (1′) = 60 seconds (60″)

The symbol ″ is used for seconds.

$$1° = 60' = 3600''$$

Standard notation requires each part of the measure to be written so that no larger measure can be made from it.

How to

Add or subtract angle measures:

1. Arrange the two measures so the like parts are aligned vertically. Seconds aligns under seconds, minutes aligns under minutes, etc.
2. Add or subtract each type of measure separately.
3. In subtraction, borrowing may be required. To borrow, take 1 unit from the next higher measure written on the left and rewrite it in terms of the needed measure.
4. Write the result in standard form.

EXAMPLE   To add $12°15'54''$ and $82°28'19''$:

$$
\begin{array}{r}
12°\,15'\quad 54'' \\
+\ 82°\,28'\quad 19'' \\
\hline
94°\,43'\quad 73'' \\
+1' - 60'' \\
\hline
94°\,44'\quad 13''
\end{array}
$$

Arrange the measures in columns of like measures and add.

Write the sum in standard notation.

Because $73'' = 1'13''$, write in standard notation by adding 1′ to the minutes column and subtracting 60″ from the seconds column. **Thus, $94°43'73'' = 94°44'13''$.**

∟

How to

Change minutes to a decimal part of a degree:

Multiply minutes by the unity ratio $\dfrac{1°}{60\ \text{min}}$.

$$\frac{x\ \text{min}}{1} \times \frac{1°}{60\ \text{min}} = \frac{x°}{60}$$

EXAMPLE   To change 15' to its decimal degree equivalent:

$$\frac{15 \text{ min}}{1} \times \frac{1°}{60 \text{ min}} = \frac{15°}{60} = 0.25°$$

⌞

How to

Change seconds to a decimal part of a degree:

Multiply seconds times two unity ratios $\dfrac{1 \text{ min}}{60 \text{ sec}}$ and $\dfrac{1°}{60 \text{ min}}$.

$$\frac{x \text{ sec}}{1} \times \frac{1 \text{ min}}{60 \text{ sec}} \times \frac{1°}{60 \text{ min}} = \frac{x°}{3{,}600}$$

EXAMPLE   To change 37″ to its decimal degree equivalent:

$$37 \text{ sec} \times \frac{1 \text{ min}}{60 \text{ sec}} \times \frac{1°}{60 \text{ min}} = \qquad \text{Multiply.}$$

$$\frac{37}{3{,}600} = 0.0103° \qquad\qquad \text{To the nearest ten-thousandth.}$$

⌞

EXAMPLE   To subtract 3°12′30″ from 15°:

$$\begin{array}{ll}
\overset{\overset{\scriptstyle 59 \quad 60}{\scriptstyle 14 \quad 60}}{\cancel{15}°} & = 14°59′60″ \qquad \text{Borrow 1° from 15°; 1° = 60′. Borrow 1′}\\
-\ \ 3°12′30″ & = \underline{\ \ 3°12′30″} \qquad \text{from 60′; 1′ = 60″, then subtract.}\\
& \quad\ \ 11°47′30″
\end{array}$$

⌞

How to

Change a decimal part of a degree to minutes:

Multiply the decimal part of a degree (x) by 60.

$$\frac{x°}{1} \times \frac{60 \text{ min}}{1°} = 60x \text{ min}$$

Polygons

EXAMPLE   To change 0.75° to minutes:

$$\frac{0.75°}{1} \times \frac{60 \text{ min}}{1°} = 0.75(60) \text{ min} = 45'$$ Multiply decimal part of degree by 60.

How to

Change a decimal part of a degree to minutes and seconds:

1. Multiply the decimal part of a degree (y) by 60 to change to minutes.

$$\frac{y \text{ deg}}{1} \times \frac{60 \text{ min}}{1 \text{ deg}} = 60y \text{ min}$$

2. Multiply the decimal part of the minute (z) by 60 to change to seconds.

$$\frac{z \text{ min}}{1} \times \frac{60 \text{ sec}}{1 \text{ min}} = 60z \text{ sec}$$

3. Write the result in minutes and seconds.

EXAMPLE   To change 0.43° to minutes and seconds:

$$\frac{0.43 \text{ deg}}{1} \times \frac{60 \text{ min}}{1 \text{ deg}} = (0.43)(60) \text{ min} = 25.8'$$ Decimal part of degree to minutes.

$$\frac{0.8 \text{ min}}{1} \times \frac{60 \text{ sec}}{1 \text{ min}} = 48''$$ Decimal part of minute to seconds. Write the result in minutes and seconds.

Thus, 0.43° = 25'48''.

How to

Multiply an angle measure by a number:

1. Multiply the number of each unit type of angle measure by the number.
2. Change units to the next larger unit of measure, where possible, so the sum is written in standard notation.

EXAMPLE An angle of 42°27′32″ needs to be increased to 4 times its size. To find the measure of the new angle in degrees, minutes, and seconds:

$$
\begin{array}{rrr}
42° & 27′ & 32″ \\
\times & & 4 \\
\hline
168° & 108′ & 128″ \\
& +2′ & -120″ \\
\hline
168° & 110′ & 8″ \\
+1° & -60′ & \\
\hline
169° & 50′ & 8″ \\
\end{array}
$$

Multiply each part of the measure by 4.

Two minutes equal 120 seconds.

One degree equals 60 minutes.
Standard notation.

How to

Divide an angle measure by a number:

1. Divide the number of the largest unit type of angle measure by the number. Write the partial quotient.
2. Multiply the partial quotient by the divisor.
3. Subtract the like units to get a remainder.
4. Change the units of the remainder to the next smaller unit.
5. Add like units.
6. Repeat the process beginning with Step 1. until all units have been divided.
7. Write the last remainder as the numerator of a fraction that has the number (divisor) as its denominator.

EXAMPLE An angle of 70°15′16″ is divided into three equal angles. To find the measure of each angle in degrees, minutes, and seconds:

$$
\begin{array}{r}
23° \quad 25′ \quad 5\tfrac{1}{3}″ \\
3\overline{)70° \quad 15′16″} \\
69° \\
\overline{1° = 60′} \\
\overline{75′} \\
75′ \\
\overline{16″} \\
15″ \\
\overline{1″}
\end{array}
$$

or **23°25′5″**   To the nearest second.

$70° \div 3 = 23°; 23° \times 3 = 69°$.
$70° - 69° = 1°; 1° \times \frac{60′}{1°} = 60′$.
$15′ + 60′ = 75′$.
$75′ \div 3 = 25′$.
$25′ \times 3 = 75′$.
$75′ - 75′ = 0$, so bring down 16″.
$16″ \div 3 = 5″; 5″ \times 3 = 15″$.
$16″ - 15″ = 1″$.
Write final remainder as numerator of fraction with 3 as denominator and add to 5″. $5″ + \tfrac{1}{3}″ = 5\tfrac{1}{3}″$.

Polygons

A **regular polygon** is a polygon with equal sides and equal angles.

A regular polygon having three sides is called an **equilateral triangle**. A regular polygon having four sides is a **square**.

How to

Find the number of degrees in each angle of a regular polygon:

Multiply the number of sides less 2 by 180° and divide by the number of sides or angles.

$$\text{Degrees per angle} = \frac{180° \ (\text{Number of sides} - 2)}{\text{Number of sides}}$$

EXAMPLE To find the number of degrees in each angle of the regular polygons:

Triangle (3 sides); Square (4 sides); Hexagon (6 sides); Octagon (8 sides)

Triangle: $\dfrac{180° \ (3 - 2)}{3} = \dfrac{180° \ (1)}{3} = 60°$

Square: $\dfrac{180° \ (4 - 2)}{4} = \dfrac{180° \ (2)}{4} = \dfrac{360°}{4} = 90°$

Hexagon: $\dfrac{180° \ (6 - 2)}{6} = \dfrac{180° \ (4)}{6} = \dfrac{720°}{6} = 120°$

Octagon: $\dfrac{180° \ (8 - 2)}{8} = \dfrac{180° \ (6)}{8} = \dfrac{1,080°}{8} = 135°$

⌞

■ 20–2 Triangles

A **scalene triangle** is a triangle with *all* three sides unequal.

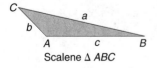

Scalene △ *ABC*

If the three sides of a triangle are unequal, the *largest* angle is opposite the *longest* side, and the *smallest* angle is opposite the *shortest* side.

An **isosceles triangle** is a triangle that has *exactly* two equal sides. The angles opposite these equal sides are also equal.

Isosceles $\triangle ABC$
$AC = BC$
$\angle A = \angle B$

An **equilateral triangle** is a triangle that has three equal sides. The three angles of an equilateral triangle are also equal. Each angle measures 60°.

Equilateral $\triangle ABC$
$AB = BC = AC$
$\angle A = \angle B = \angle C$

The **Pythagorean Theorem** states that the square of the **hypotenuse** of a right triangle is equal to the sum of the squares of the two **legs** of the triangle.

Formula for Pythagorean Theorem:

$$c^2 = a^2 + b^2$$

where c is the hypotenuse of a right triangle; a and b are the legs.

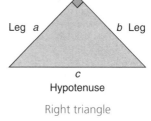

Leg a b Leg
c
Hypotenuse

Right triangle

Polygons

● **RELATED TOPIC: Distance between two points on a rectangular coordinate system**

How to

Find a leg or the hypotenuse of a right triangle:

1. Identify the two known sides of the triangle and the missing side and state the theorem symbolically.
2. Substitute known values in the Pythagorean Theorem and solve for the missing value.

EXAMPLE   To find b when $a = 8$ mm and $c = 17$ mm:

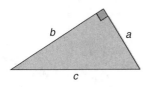

$c^2 = a^2 + b^2$ State theorem symbolically.

$17^2 = 8^2 + b^2$ Substitute known values.

$289 = 64 + b^2$

$289 - 64 = b^2$ Transpose to isolate.

$225 = b^2$

$\sqrt{225} = b$ Take the square root of both sides.

15 mm $= b$

An **isosceles right triangle** is a right triangle that has equal legs. The angles opposite the equal legs are **base angles.** The two base angles are also equal because they are the angles opposite the equal sides.

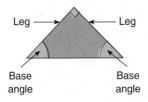

The sum of the angles of the triangle is 180°. The sum of the two base angles of a right triangle is 180° − 90° or 90°. If both angles are equal, as in the isosceles right triangle, then each angle is $\frac{1}{2}(90°)$ or 45°. Thus, we get the name 45°, 45°, 90° triangle.

A **45°, 45°, 90° triangle** is an isosceles right triangle and the two base angles are 45° each.

How to

Find the hypotenuse of a 45°, 45°, 90° triangle:

1. Identify the known and unknown parts of the triangle and substitute in the formula: Hypotenuse = Leg$\sqrt{2}$.
2. Solve for the missing value.
3. Write in simplest radical form or evaluate as a decimal value.

EXAMPLE   To find AC and BC if $AB = 5$ cm:

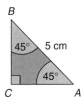

$$\text{Hypotenuse} = \text{Leg}\sqrt{2}$$ The hypotenuse is equal to the product of a leg and $\sqrt{2}$.

$$5 = AC\sqrt{2}$$ Solve for AC, a leg. Divide both sides by $\sqrt{2}$.

$$\frac{5}{\sqrt{2}} = \frac{AC\sqrt{2}}{\sqrt{2}}$$

$$\frac{5}{\sqrt{2}} = AC$$ Rationalize denominator. Multiply numerator and denominator by 1 in the form of $\frac{\sqrt{2}}{\sqrt{2}}$.

$$\frac{5}{\sqrt{2}} \cdot \frac{\sqrt{2}}{\sqrt{2}} = AC$$

$$\frac{5\sqrt{2}}{2} = AC$$ Replace $\sqrt{2}$ with decimal approximation and simplify.

$$3.536 \text{ cm} = AC$$ Rounded.

Using a calculator, we could have found the decimal value of AC without first rationalizing the denominator.

$$AC = \frac{5}{\sqrt{2}}$$

$AC = 3.536$ cm, from $\boxed{5}$ $\boxed{\div}$ $\boxed{\sqrt{}}$ $\boxed{2}$ $\boxed{=}$ $\Rightarrow$ 3.536   Rounded.

Since $AC = BC$ **then** $BC = 3.536$ **cm.**

⌞

● **RELATED TOPIC: Rationalize denominators**

The **altitude of an equilateral triangle** is a line drawn from the midpoint of the base to the opposite vertex, dividing the triangle into two congruent right triangles.

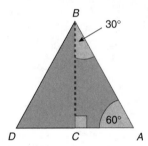

Because the altitude of an equilateral triangle bisects (divides in half) the vertex angle and the base, two 30° angles are formed at B and $AC = CD$. We can also say that $AC = \frac{1}{2}AD$ or one-half of any side of the original equilateral triangle. That means $AC = \frac{1}{2}AB$, or the side opposite the 30° angle is half the hypotenuse.

A **30°, 60°, 90° right triangle** is a special right triangle with angles that measure 30°, 60°, and 90°, respectively, and a hypotenuse that is twice the side opposite the 30° angle. The side (leg) opposite the 60° angle is equal to $\sqrt{3}$ times the side opposite the 30° angle.

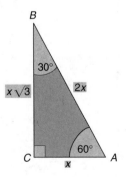

How to

Find the missing sides of a 30°, 60°, 90° triangle if the side opposite the 30° angle is known:

1. Find the hypotenuse by multiplying the known side by 2.
2. Find the side opposite the 60° angle by multiplying the known side by $\sqrt{3}$.

EXAMPLE   To find *AB* and *BC* if *AC* = 7 inches:

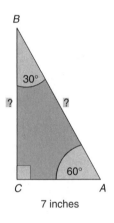

$AB = 2 \times 7$ inches = **14 inches**

$BC = 7$ inches $\times \sqrt{3} = 7\sqrt{3}$ **inches**

How to

Find the missing sides of a 30°, 60°, 90° triangle if the side opposite the 30° angle is unknown:

1. Use the appropriate relationship between the known side and the side opposite the 30° angle to find the side opposite the 30° angle.
2. Use the procedure in the previous "How to" box to find the other missing side (hypotenuse or side opposite the 60° angle).

Polygons

EXAMPLE To find AC and AB when BC = 8 cm:

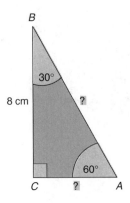

Find AC:

$$BC = AC\sqrt{3} \qquad \text{Let } x = AC, BC = 8.$$

$$8 = x\sqrt{3} \qquad \text{Solve for } x.$$

$$\frac{8}{\sqrt{3}} = \frac{x\sqrt{3}}{\sqrt{3}} \qquad \text{Divide both sides by } \sqrt{3}.$$

$$\frac{8}{\sqrt{3}} = x \qquad \text{Rationalize the denominator.}$$

$$x = \frac{8}{\sqrt{3}} \cdot \frac{\sqrt{3}}{\sqrt{3}}$$

$$x = \frac{8\sqrt{3}}{3} \text{ cm}$$

$$AC = \frac{8\sqrt{3}}{3} \text{ cm (exact)   or   4.6 (approximate)}$$

Find AB:

$$AB = 2AC \qquad \text{Substitute } \tfrac{8\sqrt{3}}{3} \text{ cm for } AC.$$

$$AB = 2\left(\frac{8\sqrt{3}}{3}\right)$$

$$AB = \frac{16\sqrt{3}}{3} \text{ cm (exact)   or   9.2 (approximate)}$$

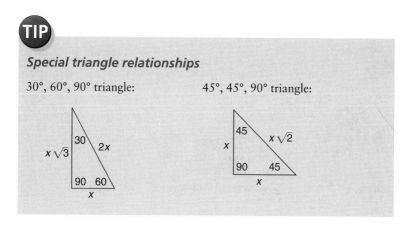

TIP

Special triangle relationships

30°, 60°, 90° triangle: 45°, 45°, 90° triangle:

■ 20–3 Inscribed and Circumscribed Regular Polygons and Circles

A polygon is **inscribed in a circle** if it is inside the circle and all its **vertices** (points where sides of each angle meet) are on the circle. The circle is said to be **circumscribed about the polygon.**

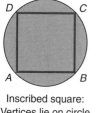

Inscribed square:
Vertices lie on circle
at points *A*, *B*, *C*, and *D*.

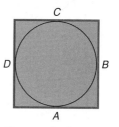

Circumscribed square:
Sides of square are tangent
at points *A*, *B*, *C*, and *D*.

Polygons

A polygon is **circumscribed about a circle** if it is outside the circle and all its sides are **tangent** to (intersecting in exactly one point) the circumference. The circle is said to be **inscribed in the polygon.**

When an equilateral triangle is inscribed in a circle, the radius (from the vertex of the triangle to the center of the circle) is two-thirds the height of the triangle. When the equilateral triangle is circumscribed about a circle, the radius (from the center of the circle to the base of the triangle) is one-third the height of the triangle.

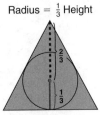

Radius = $\frac{2}{3}$ Height   Radius = $\frac{1}{3}$ Height

Inscribed triangle   Circumscribed triangle

EXAMPLE To find the dimensions for the inscribed equilateral triangle where $CD = 13$ mm and $AO = 15$ mm:

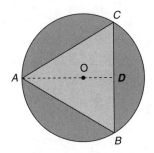

(a) Height of $\triangle ABC$ (b) BC (c) BD (d) $\angle CAD$

(e) $\angle ACD$ (f) Area of $\triangle ABC$

(a) $AO = \dfrac{2}{3}AD$ $AO = \frac{2}{3}$ of height of AD. Substitute.

         $15 = \dfrac{2}{3}AD$ Solve for AD.

       $\dfrac{3}{2}(15) = \left(\dfrac{3}{2}\right)\dfrac{2}{3}AD$

         $\dfrac{45}{2} = AD$

     $22.5 \text{ mm} = AD$

Or use the formula for finding a leg of a 30°, 60°, 90° triangle:

$$AD = CD\sqrt{3}$$

$$AD = 13\sqrt{3}$$

$$AD = 22.5 \text{ mm}$$ Rounded.

(b) $BC = 2CD$ 30°, 60°, 90° △

    $BC = 2(13)$

    $BC = 26 \text{ mm}$

(c) $BD = 13 \text{ mm}$ AD bisects BC.

(d) $\angle CAD = 30°$ $\frac{1}{2}(60°)$

(e) $\angle ACD = 60°$

(f) $A = \dfrac{1}{2}bh$

    $A = \dfrac{1}{2}(BC)(AD)$

    $A = \dfrac{1}{2}(26)(22.5)$

    $A = 292.5 \text{ mm}^2$

Diagonals of a square:

The diagonals of a square form congruent 45°, 45°, 90° triangles.

Diameter of a circle inscribed in a square:

The diameter of a circle inscribed in a square equals a side of the square.

Diameter of a circle circumscribed about a square:

The diameter of a circle circumscribed about a square equals a diagonal of the square.

Polygons

Radius of a circle inscribed in a square:

The radius of a circle inscribed in a square equals the height of a 45°, 45°, 90° triangle formed by two diagonals.

Radius of a circle circumscribed about a square:

The radius of a circle circumscribed about a square equals the height of a 45°, 45°, 90° triangle formed by one diagonal.

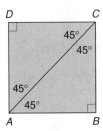

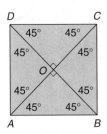

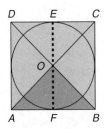

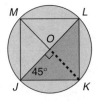

How to

Use the properties of inscribed and circumscribed squares to find missing amounts:

1. Determine how the known value can be used with the relationships of a circumscribed or inscribed square to find the missing amount.
2. Use properties of a right triangle (Pythagorean Theorem or special right triangle) or circle (radius or diameter) to calculate the missing value.

▲ APPLICATION: Measurements for Making Furniture

EXAMPLE   To find the minimum depth of cut required to mill a circle on the end of a square piece of stock with a side measuring 4 cm:

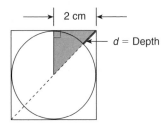

The side of the square is 4 cm, so the diameter of the circle is 4 cm and the radius is 2 cm. If the radius is drawn perpendicular to the top side, a right triangle is formed as indicated (shading).

Let d = depth of milling, which equals the hypotenuse of the shaded triangle minus the radius of the circle. Use the Pythagorean Theorem, to find the hypotenuse where c = hypotenuse.

$c^2 = a^2 + b^2$ Substitute known values.

$c^2 = 2^2 + 2^2$

$c^2 = 4 + 4$

$c^2 = 8$

$c = \sqrt{8}$ Hypotenuse of triangle.

$c = 2.828427125$ cm

Hypotenuse − radius = depth of milling.

$2.828427125 - 2 = 0.83$ cm   Rounded.

The minimum depth of milling is 0.83 cm.

└

Polygons

A **regular hexagon** is a figure with six equal sides and six equal angles of 120° each. Three diagonals joining pairs of opposite vertices divide the hexagon into six congruent equilateral triangles (all angles 60°).

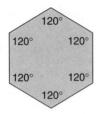

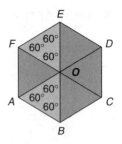

Diagonals join opposite vertices of a regular hexagon:

The three diagonals that join opposite vertices of a regular hexagon form six congruent equilateral triangles.

In the inscribed circle, the height of an equilateral triangle (GO) is the radius, and for the circumscribed circle, any side of the equilateral triangles (AO) equals the radius. The height divides any of the six triangles into two congruent 30°, 60°, 90° right triangles.

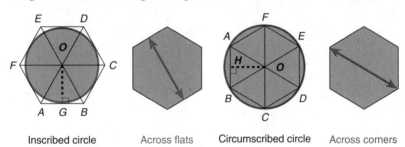

| Inscribed circle | Across flats | Circumscribed circle | Across corners |

Radius of a circle inscribed in a regular hexagon:

The radius of a circle inscribed in a regular hexagon is the height of an equilateral triangle formed by the three diagonals.

Distance across flats:

The diameter of a circle inscribed in a regular hexagon is the **distance across flats** or twice the height of any equilateral triangle formed by the three diagonals.

Radius of a circumscribed circle:

The radius of a circle circumscribed about a regular hexagon is a side of an equilateral triangle formed by the three diagonals or one-half the distance across corners.

Distance across corners:

The diameter of a circle circumscribed about a regular hexagon is a diagonal of the hexagon (**distance across corners**).

EXAMPLE   To find the indicated dimensions using the first figure for (a)–(f) and the second figure for (g)–(j) (when appropriate, round to hundredths):

(a) $\angle EOD$ (b) $\angle COH$ (c) $\angle BHO$ (d)   Radius

(e)   Distance across corners   (f) FO (g)   Distance across flats

(h) MO (i)   Diagonal   (j) $\angle LMN$

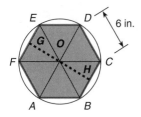
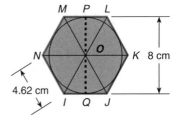

(a) $\angle EOD = 60°$ This is an angle of an equilateral triangle.

(b) $\angle COH = 30°$ Height bisects the 60° angle to form a 30°, 60°, 90° triangle.

(c) $\angle BHO = 90°$ The height forms a right angle with the base.

(d)   Radius = 6 in.   The radius is a side of the equilateral triangle.

(e)   Distance across corners = 12 in.   The distance across corners is the diameter.

(f) $FO = 6$ in. This is a side of an equilateral triangle.

(g) Distance across flats $= 8$ cm This is the diameter of the inscribed circle.

(h) $MO = 4.62$ cm An equilateral triangle has equal sides.

(i) Diagonal $= 9.24$ cm The diagonal is twice the side of the equilateral triangle.

(j) $\angle LMN = 120°$ This is the angle at the vertex of a hexagon.

└─

Statistics and Probability

CHAPTER 21

Interpreting and Analyzing Data

- 21–1 **Reading Circle, Bar, and Line Graphs**
- 21–2 **Frequency Distributions, Histograms, and Frequency Polygons**
- 21–3 **Finding Statistical Measures**
- 21–4 **Counting Techniques and Simple Probability**

■ 21–1 Reading Circle, Bar, and Line Graphs

A **circle graph** uses a divided circle to show pictorially how a total amount is divided into parts.

A **bar graph** uses two or more bars to compare two or more amounts. The bar lengths represent the amounts being compared. Bars can be drawn either horizontally or vertically.

The **axis** or **reference line** that runs along the length of the bars is a scale of the amounts being compared. The other reference line labels the bars.

A **line graph** uses one or more lines to show changes or trends in data.

The **horizontal axis** or **reference line** on a line graph usually represents periods of time or specific times. The **vertical axis** or reference line is usually scaled to represent numerical amounts. Line graphs show trends in data and high values and low values at a glance.

How to

Read a circle, bar, or line graph:

1. Examine the title of the graph to find out what information is shown.
2. Examine the parts to see how they relate to one another and to the whole.

3. Examine the labels for each part of the graph and any explanatory remarks that may be given.

4. Use the given parts to calculate additional amounts.

EXAMPLE   To answer the questions from the circle graph:

  (a)   What percent of the wholesale price is the cost of labor?

  (b)   What percent of the wholesale price is the cost of materials?

  (c)   What would the wholesale price be if no tariff (tax) was paid on imported parts?

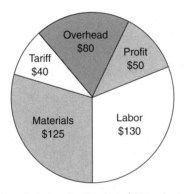

Distribution of wholesale price for a $425 color television.

(a) $$\frac{R}{100} = \frac{130}{425}$$ R is the percent of wholesale price that is labor cost.

$$R \times 425 = 13{,}000$$

$$R = \frac{13{,}000}{425}$$

$$R = 30.58823529$$

$$R = 30.6\% \text{ (labor)}$$

(b) $$\frac{R}{100} = \frac{125}{425}$$ R is the percent of wholesale price that is materials cost.

$$R \times 425 = 12{,}500$$

$$R = \frac{12,500}{425}$$

$$R = 29.41176471$$

$$R = 29.4\% \text{ (materials)}$$

(c)   Price − tariff = 425 − 40 = $385 (cost without tariff)

⌞

● **RELATED TOPIC: Proportions**

EXAMPLE   To answer the questions using the bar graph:

(a)   How many more 100 million barrels of oil are indicated for the company in 2000 than in 1980?

(b)   Judging from the graph, should company oil production in 2005 be more or less than in 2000?

(c)   How many 100 million barrels of oil did the company produce in 1990?

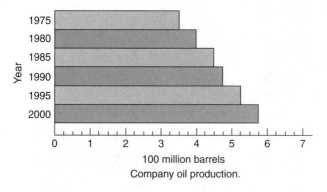

100 million barrels
Company oil production.

(a)   5.75 − 4 = **1.75**

(b)   **More. The number of barrels increases each year.**

(c)   **4.75**

⌞

EXAMPLE   To answer the questions regarding a patient's temperature using the graph:

(a)   On what date and time of day did the patient's temperature first drop to within 0.2° of normal (98.6°F)?

(b) On which post-op (postoperative) days was the patient's temperature within 0.2° of normal?

(c) What was the highest temperature recorded for the patient?

Date	4-10-2003						4-11-2003						4-12-2003					
Weight																		
Post-op Day	1						2						3					
	A.M.			P.M.			A.M.			P.M.			A.M.			P.M.		
F	12	4	8	12	4	8	12	4	8	12	4	8	12	4	8	12	4	8

Time

Graphic temperature chart.

(a) **4-11-03 at 4 A.M.** (Each "dot" is 0.2°, so the temperature was 98.8°F.)

(b) **Post-op days 2 and 3** (beginning at 4 A.M. on day 2)

(c) **102.2°** (recorded at 12 A.M. on 4-10-03)

■ 21–2 Frequency Distributions, Histograms, and Frequency Polygons

Suppose for a class of 25 students the instructor records the following grades:

76 91 71 83 97 87 77 88 93 77 93 81 63

79 74 77 76 97 87 89 68 90 84 88 91

The scores can be arranged into several smaller groups, called **class intervals**. The word **class** means a special category.

These scores can be grouped into class intervals of 5, such as 60–64, 65–69, 70–74, 75–79, 80–84, 85–89, 90–94, and 95–99. Each class interval has an odd number of scores. The **middle score** of each interval is a **class midpoint**.

A **tally** or count of the number of scores that fall into each class interval is a **class frequency**.

A compilation of class intervals, midpoints, tallies, and class frequencies is called a **grouped frequency distribution**.

EXAMPLE   To use the grouped frequency distribution to determine the information:

(a)   The number of students that scored 70 or above:

$2 + 6 + 3 + 5 + 5 + 2 = 23$ Add the frequencies for classes with scores 70 or higher.

23 students scored 70 or above.

(b)   The number of students that made A's (90 or higher):

$5 + 2 = 7$

7 students made A's (90 or higher).

(c)   The percent of the total grades that were A's (90's):

$$\frac{7 \text{ A's}}{25 \text{ total}} = \frac{7}{25} = 0.28 = 28\% \text{ A's}$$

Grouped Frequency Distribution of 25 Scores

Class Interval	Midpoint	Tally	Class Frequency
60–64	62	/	1
65–69	67	/	1
70–74	72	//	2
75–79	77	## /	6
80–84	82	///	3
85–89	87	##	5
90–94	92	##	5
95–99	97	//	2

A **histogram** is a bar graph that uses two scales, one for class intervals and one for class frequencies.

Make a histogram for a data set:

1. Create a table (frequency distribution) with headings *Class Interval, Midpoint, Tally, Class Frequency.*
2. Use the data to complete the frequency distribution.
3. Show the class intervals on one axis of a bar graph with spacing to provide for the width of the bars. Increment the other axis based on the numbers shown in the class frequency column.
4. Use the class frequency data to draw a bar for each of the class intervals.
5. Label both axes to reflect the information being displayed. Provide a title for the histogram.

EXAMPLE   Students in a history class reported their credit-hour loads: 3, 12, 15, 3, 6, 6, 12, 9, 12, 9, 6, 3, 12, 18, 6, 9. To make a histogram:

To establish a class interval with an easy-to-find midpoint, use an odd number of points in the interval. Here, an interval of 5 is used; that is, 0–4 contains five possibilities: 0, 1, 2, 3, and 4. The middle number is the midpoint, 2. Make a tally mark for each time the credit hours of a student falls in the interval. Then count the tally marks to get the class frequency.

Frequency Distribution of Credit-Hour Loads				
Class Interval	Midpoint	Tally	Class Frequency	
0–4	2	///	3	Create a frequency distribution from given data.
5–9	7	//// //	7	
10–14	12	////	4	
15–19	17	//	2	

Statistics
Probability

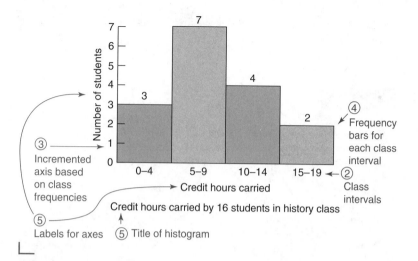

Use a histogram to provide desired information:

1. Determine the information that is desired.
2. Read from the histogram directly information regarding frequencies for particular class intervals or vice versa.
3. Use mathematical operations and data from the histogram to calculate information not provided directly from the histogram.

EXAMPLE   To use the histogram for credit hours earned to find:

(a)   The number of students that carried 5–9 credit hours.

The bar for 5–9 credit hours shows 7 students.

(b)   The number of students that carried less than 15 hours.

$3 + 7 + 4 = $ **14 students**   Three class intervals show students that carried less than 15 hours, so we add.

(c)   The percent of students that carried 10–14 credit hours.

$$\frac{4}{16} = \frac{1}{4} = 0.25 = 25\%$$ Since 4 students carried 10–14 credit hours and there are 16 students (scores) in the study, the percent is calculated using 16 as the base and 4 as the percentage.

A **frequency polygon** is a line graph that joins the midpoints of the bars of a histogram and begins with the beginning point of the first bar and ends with the ending point of the last bar.

How to

Make a frequency polygon:

1. Begin with a histogram and identify each class midpoint with a dot (point) on the top of each bar.
2. Connect the points to make a line graph.
3. Remove the bars and connect the end points of the line graph from Step 2 to the horizontal axis at the smallest value of the first interval and the largest value of the last interval on the right.

EXAMPLE   To make a frequency polygon using the data in the frequency distribution that illustrates the number of years of service for 20 employees:

Grouped Frequency Distribution of 20 Employees			
Class Interval	Midpoint	Tally	Class Frequency
0–2	1	_HH III_	8
3–5	4	_HH I_	6
6–8	7	_IIII_	4
9–11	10	_II_	2

Years of service of 20 employees

Make a histogram from the frequency distribution.

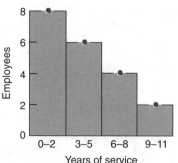

Identify each class midpoint with a point.

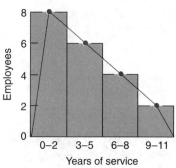

Make a line graph. Connect 0 and 11.

Horizontal scale aligns with the midpoints of each class interval. Remove bars.

EXAMPLE   To use the frequency polygon to answer the questions:

(a)   What can you say about the relationship between the years of service and the number of employees?

As the number of years of service increases the number of employees decreases.

(b) How many employees have an average of 4 years of service?

6 employees

(c) What is the average number of years of service for employees in the 6 to 8 years of service bracket?

7 is the midpoint or average for the interval.

■ 21–3 Finding Statistical Measures

An **average** is an approximate number that is a central value or typical value of a group of data values. The most common average is the arithmetic mean or arithmetic average. The **arithmetic mean** or **mean** is the sum of the quantities in the data set divided by the number of quantities.

Symbols can be used to write the procedures for statistical measures. The formula in symbols for the mean for a set of data is

$$\bar{x} = \frac{\Sigma x_i}{n}$$

The formula is read as "the mean $\bar{x}$ (read x-bar) equals the sum Σ of each value of x (read x sub i) divided by the number of values (n)." The Greek capital letter **sigma**, Σ, is a summation symbol and indicates the addition of a group of values. The notation x_i identifies each data value with a subscript: x_1 is the first value, x_2 is the second value, and x_n is the nth value.

How to

Find the arithmetic mean or arithmetic average:

Find the mean of 22, 31, and 37.

$$x_1 = 22, x_2 = 31, x_3 = 37$$

1. Add the quantities. Σx_i $\Sigma x_i = 22 + 31 + 37$

$$\Sigma x_i = 90$$

2. Divide the sum by the number of quantities.

    Symbolically, $\bar{x} = \dfrac{\Sigma x_i}{n}$

$$\bar{x} = \frac{\Sigma x_i}{n} = \frac{90}{3}$$

$$\bar{x} = 30$$

Statistics Probability

EXAMPLE To find the mean of each group of quantities:

(a) Pulse rates: 68, 84, 76, 72, 80
There are 5 pulse rates, so we find the sum and divide by 5:

$$\bar{x} = \frac{\Sigma x_i}{n} = \frac{68 + 84 + 76 + 72 + 80}{5} = \frac{380}{5} = 76$$

The mean pulse rate is 76.

(b) Pounds: 21, 33, 12.5, 35.2 (to the nearest hundredth)
There are 4 weights, so we find their sum and divide by 4.

$$\bar{x} = \frac{\Sigma x_i}{n} = \frac{21 + 33 + 12.5 + 35.2}{4} = \frac{101.7}{4} = 25.425$$

The mean weight is 25.43 lb (to nearest hundredth).

The **median** is the middle value of a set of data values arranged in order of size. The median is frequently used for data relating to annual incomes, housing price ranges, etc.

How to

Find the median:

1. Arrange the values in order of size.
2. If the number of values is odd, the median is the middle value.
3. If the number of values is even, the median is the average of the two middle values.

EXAMPLE A TPR chart shows a patient's temperature, pulse rate, and respiration rate. The following pulse rates were recorded on a TPR chart: 68, 88, 76, 64, 72. To find the patient's median pulse rate:

88   Arrange data in order of size.
76
72 ←   The median is the middle value in an odd number of values.
68
64

The median pulse rate is 72.

EXAMPLE   The following temperatures were recorded: 56°, 48°, 66°, and 62°. To find the median temperature:

48   Arrange data in order of size.

$\left. \begin{array}{c} 56 \\ 62 \end{array} \right\} \dfrac{56 + 62}{2} = \dfrac{118}{2} = 59$ The number of temperatures is even; the median is the average of the two middle values.

66

The median temperature is 59°.

└

The **mode** is the most frequently occurring value in the data set.

How to

Find the mode:

1. Identify the value or values that occur with the greatest frequency as the mode.
2. If no value occurs more than another value, there is *no mode* for the data set.
3. If more than one value occurs with the greatest frequency, the modes of the data set are the values that have the greatest frequency.

EXAMPLE   The hourly pay rates at a local fast-food restaurant are: cooks, $7.50; servers, $7.15; bussers, $7.15; dishwashers, $9.25; managers, $10.50. To find the mode:

Identify the value or values that appear most often.

The hourly pay rate of $7.15 occurs more than any other rate. It is the mode.

└

How to

Find the range of a set of data:

1. Find the highest and lowest values.
2. Find the difference between the highest and lowest values.

EXAMPLE   To find the range for the data described in the example for fast-food restaurant hourly pay rates:

The high value is $10.50. The low value is $7.15.

Range = $10.50 − $7.15 = **$3.35**

└

Use more than one statistical measure

A common mistake when making conclusions or inferences from statistical measures is to examine only one statistic, such as the mean. To obtain a complete picture of the data requires looking at more than one statistic.

The range gives some information about dispersion, but does not specify whether the highest or lowest values are typical values or extreme outliers. A clearer picture of the data set is obtained by examining how much each data point *differs* or *deviates* from the mean. The **deviation from the mean** of a data value is the difference between the value and the mean.

How to

Find the deviations from the mean of a set of data:

1. Find the mean of a set of data.

$$\bar{x} = \frac{\text{Sum of data values}}{\text{Number of values}} = \frac{\Sigma x_i}{n}$$

Data set: 38, 43, 45, 44

$$\frac{38 + 43 + 45 + 44}{4} =$$

$$\frac{170}{4} = 42.5$$

2. Find the amount by which each data value deviates or is different from the mean.

$$\text{Deviation} = \frac{\text{Data}}{\text{value}} - \text{Mean}$$

$$= x_i - \bar{x}$$

$38 - 42.5 = -4.5$
(below the mean)
$43 - 42.5 = 0.5$
(above the mean)
$45 - 42.5 = 2.5$
(above the mean)
$44 - 42.5 = 1.5$
(above the mean)

When the value is smaller than the mean, the difference is represented by a *negative* number indicating the value is *below* or less than the mean. When the value is larger than the mean, the difference is represented by a *positive* number indicating the value is *above* or greater than the mean. *The absolute value of the sum of the deviations below the mean should equal the sum of the deviations above the mean.* In the example in the box, only one value is below the mean, and its deviation is −4.5. Three values are above the mean, and the sum of these deviations is 0.5 + 2.5 + 1.5 = 4.5. *The sum of all deviations from the mean for any set of data is zero.*

EXAMPLE   To find the deviations from the mean for the set of data 45, 63, 87, and 91:

$$\bar{x} = \frac{45 + 63 + 87 + 91}{4} = \frac{286}{4} = 71.5$$ Find the mean of the data set.

Values i	x_i	Deviations $x_i - \bar{x}$	
1	45	45 − 71.5 = −26.5	A table is a good way to
2	63	63 − 71.5 = −8.5	organize the calculation
3	87	87 − 71.5 = 15.5	of the deviations from
4	91	91 − 71.5 = 19.5	the mean.
Total	286	0	

Since the sum of the deviations from the mean is zero, statisticians employ a statistical measure called the *standard deviation,* which uses the square of each deviation from the mean. The square of a negative number is always positive. The squared deviations are averaged and the result is called the **variance.** The square root of the variance is called the **standard deviation.** Other formulas exist for finding the standard deviation of a set of values, but only one is included here. This formula requires several calculations that are best organized in a table.

$$s = \sqrt{\frac{\Sigma(x_i - \bar{x})^2}{n - 1}}$$ s = standard deviation

How to

Find the standard deviation of a set of data:

1. Find the mean, $\bar{x}$.
2. Find the deviation of each value from the mean: $(x_i - \bar{x})$.
3. Square each deviation: $(x_i - \bar{x})^2$.
4. Find the sum of the squared deviations: $\Sigma(x_i - \bar{x})^2$.

Statistics
Probability

5. Divide the sum of the squared deviations by *one less than* the number of values in the data set. This amount is the *variance, v.*

$$v = \frac{\Sigma(x_i - \bar{x})^2}{n - 1}$$

6. Find the square root of the variance. This amount is the *standard deviation, s.*

$$s = \sqrt{\frac{\Sigma(x_i - \bar{x})^2}{n - 1}}$$

The standard deviation formula used here deals with a *sample* of an entire population or a relatively small set of data. In this formula, instead of dividing by n, the number of data values in the sample, we divide by $n - 1$. In very large sets of data, there is very little difference in the calculations using n versus $n - 1$. The logical basis for using $n - 1$ assumes that one data value is *exactly* the **mean** (which may or may not be true) and there are $n - 1$ data values that deviate from the mean.

EXAMPLE   To find the standard deviation for the values 45, 63, 87, and 91:

$$\frac{45 + 63 + 87 + 91}{4} = 71.5 \quad \text{Find the mean.}$$

i	x_i	$x_i - \bar{x}$ Deviations from the mean.	$(x_i - \bar{x})^2$ Squares of the deviations from the mean.
1	45	$45 - 71.5 = -26.5$	$(-26.5)^2 = 702.25$
2	63	$63 - 71.5 = -8.5$	$(-8.5)^2 = 72.25$
3	87	$87 - 71.5 = 15.5$	$(15.5)^2 = 240.25$
4	91	$91 - 71.5 = 19.5$	$(19.5)^2 = 380.25$
Σ	286	0 Sum of deviations.	1,395 Sum of squared deviations.

$$v = \frac{\Sigma(x_i - \bar{x})^2}{n - 1} \quad \text{Take the square root of the variance.}$$

$$s = \sqrt{\frac{1,395}{3}} = \sqrt{465} = 21.56385865 \text{ or } \mathbf{21.6 \text{ standard deviation}}$$

To interpret the meaning of standard deviation, note that a smaller value indicates the mean is a typical value in the data set. A large standard deviation indicates that the mean is not typical, and other statistical measures should be examined to better understand the characteristics of the data set. In this example, the mean of 71.5 with a standard deviation of 21.6 indicates that the mean is not a typical value and the data values do not cluster closely around the mean.

■ 21–4 Counting Techniques and Simple Probability

A **set** is a well-defined group of objects or **elements.** The numbers 2, 4, 6, 8, and 10 can be a set of even numbers between 1 and 12. Women, men, and children can be a set of people. A, B, and C can be a set of the first three capital letters in the alphabet.

Counting, in this section, means determining all the possible ways the elements in a set can be arranged. One way to count is to *list* all possible arrangements and then count the number of arrangements.

● **RELATED TOPIC: Set notation**

EXAMPLE   To list and count the ways the elements *A*, *B*, and *C* in the set can be arranged:

A is first,	*B* is first,	*C* is first,
2 choices	2 choices	2 choices
ABC	*BAC*	*CAB*
ACB	*BCA*	*CBA*

A, B, and C can be arranged in six ways. Each of these arrangements can also be called a set.

∟

If more than three elements are in the set, it is helpful to use a **tree diagram,** which allows each new set of possibilities to branch out from a previous possibility.

How to

Make a tree diagram of possible arrangements of items in a set:

1. List the choices for putting an item in the first slot.
2. From each choice, list the remaining choices for the second slot (first branch).

3. From the end of each branch, make a branch to the remaining choices for the third slot.
4. Continue the process until there are no remaining choices.
5. To list an arrangement, start with the first slot and follow a branch to its end and record the choices.
6. To count the total number of possible arrangements, count the number of branch ends in the last slot.

EXAMPLE   To make a tree diagram and count the number of ways the elements in the set containing letters W, X, Y, and Z can be arranged:

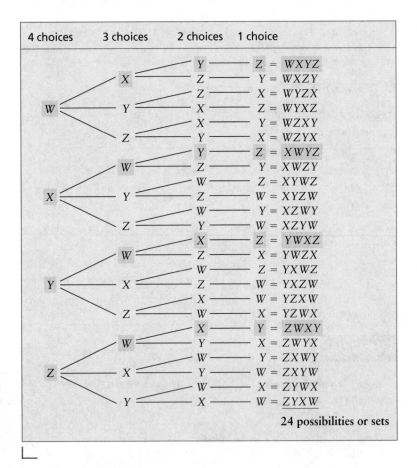

4 choices	3 choices	2 choices	1 choice

24 possibilities or sets

As evident in the previous example, the greater the number of elements in a set, the greater the complexity and time required to list all possible arrangements. We can use logic and common sense to obtain a count of the possible arrangements of a set of elements.

There are four possibilities for the first letter, *W*, *X*, *Y*, or *Z*. For each of the four possible first letters, we have three choices for second letters or 12 possibilities. For each of the 12 possibilities two choices remain for the third letter: $12 \times 2 = 24$. Then, for each of the 24 three-letter combinations, only 1 choice is left: $24 \times 1 = 24$. So, there are 24 possibilities.

How to

Determine the number of choices for arranging a specified number of items:

1. Determine the number of slots to be filled.
2. Determine the number of choices for each slot.
3. Multiply the numbers from Steps 1 and 2.

EXAMPLE A coin is tossed three times. With each toss, the coin falls heads up or tails up. To find the number of combinations of heads and tails there are with three tosses of the coin:

There are three tosses. Each toss has only two possibilities, heads or tails; that is,

1st toss	2nd toss	3rd toss
2 possibilities	2 possibilities	2 possibilities

By multiplying the number of possible outcomes for each toss, we get $2 \cdot 2 \cdot 2 = 8$. So 8 combinations are possible.

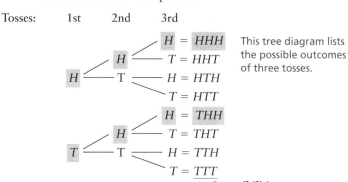

This tree diagram lists the possible outcomes of three tosses.

8 possibilities or sets

Probability means the chance of an event occurring if an activity is repeated over and over. The probability of an event occurring is expressed as a ratio or a percent.

Weather forecasters use percents when they forecast a 60% chance of rain or a 20% chance of snow.

When a coin is tossed, two outcomes are possible, heads or tails. But only one side will be up. The probability of tossing heads is 1 out of 2, or $\frac{1}{2}$.

How to

Express the probability of an event occurring successfully:

1. Determine the total number of elements in the set.
2. Determine the number of elements that are defined as successful.

3. Make a ratio: $\dfrac{\text{number of successful possibilities}}{\text{number of total possibilities}}$

4. Express the ratio in lowest terms.

EXAMPLE To determine the probability that a 3 will appear in one roll of a die:

A die has **six** sides where dots represent 1–6. The number showing on each side is an element in the set of six sides, so there are 6 elements in the set.

Only **one** element is a 3.

The probability of rolling a 3 is $\frac{1}{6}$.

Appendix

A–1 Symbols

Symbol	Meaning
$+$	Add
$-$	Subtract
$\times, \cdot, *, (\ )(\ )$	Multiply
$\div, \lceil\ , /, —$	Divide
$=$	Equal to
$\approx$	Approximately equal to
$\neq$	Not equal to
$\%$	Percent
$>$	Greater than
$<$	Less than
$\geq$	Greater than or equal to
$\leq$	Less than or equal to
$\sqrt{\ }$	Radical sign or square root
$(\ ), [\ ], \{\ \}, —$	Grouping symbols
$\|\ \|$	Absolute value

$f(x)$	Function notation, read "f of x"
$\overleftrightarrow{AB}$	Line AB
$\overline{AB}$	Line segment AB
$\overrightarrow{AB}$	Ray AB
$\approx$, $\cong$	Congruent to
$\sim$	Similar to (geometric figures)
$\angle$	Angle
$\perp$	Perpendicular
$\triangle$	Triangle
$\bigcirc$	Circle
⌐	Right angle
Δ	Delta, change, used with slope
$\{\dots \mid \dots\}$	Such that, used with set notation
$\sum$	Summation
x_1	Subscript (1)
$\{\ \}$, $\varnothing$	Empty or null set
$\in$	Is an element of
$\cup$	Union (of sets)
$\cap$	Intersection (of sets)
π	Constant—Pi (ratio of diameter to circumference of a circle, approximately 3.14)
e	Constant—natural exponential (from $\left(1 + \dfrac{1}{n}\right)^n$ where $n \to \infty$, approximately 2.72)
i	The square root of -1 $\left(\sqrt{-1}\right)$
∞	Infinity
$\therefore$	Therefore
$\exists$	There exists
$\forall$	For every

■ A–2 Roman Numerals

I	1	VIII	8
II	2	IX	9
III	3	X	10
IV	4	XI	11
V	5	XII	12
VI	6	XIII	13
VII	7	XIV	14

XV	15	CM	900
XVI	16	M	1000
XVII	17	MC	1100
XVIII	18	MCL	1150
XIX	19	MCC	1200
XX	20	MCCL	1250
XXV	25	MCCC	1300
XXX	30	MCCCL	1350
XXXV	35	MCD	1400
XL	40	MCDL	1450
L	50	MD	1500
LX	60	MDL	1550
LXX	70	MDC	1600
LXXX	80	MDCC	1700
XC	90	MDCCC	1800
C	100	MCM	1900
CX	110	MM	2000
CL	150	MMM	3000
CC	200	$M\overline{V}$	4000
CCC	300	$\overline{V}$	5000
CD	400	$\overline{X}$	10,000
D	500	$\overline{L}$	50,000
DC	600	$\overline{C}$	100,000
DCC	700	$\overline{M}$	1,000,000
DCCC	800		

■ A–3 Greek Alphabet

Many of the Greek alphabet symbols are used in mathematics. For example, σ is the letter used to indicate the population standard deviation in statistics.

A, α	alpha		H, η	eta
B, β	beta		Θ, θ	theta
Γ, γ	gamma		I, ι	iota
Δ, δ	delta		K, κ	kappa
E, ϵ	epsilon		Λ, λ	lamda
Z, ζ	zeta		M, μ	mu

N, ν	nu	T, τ	tau
Ξ, ξ	xi	Υ, υ	upsilon
O, o	omikron (or omicron)	Φ, ϕ, φ	phi
Π, π	pi	X, χ	chi
P, ρ	rho	Ψ, ψ	psi
Σ, σ, ς	sigma*	Ω, ω	omega

*Sigma has two lower-case forms. σ is used everywhere except at the end of a word. ς is used at the end of a word.

■ A–4 Formulas*

Square

Perimeter = **4s**, where s = length of a side of the square, 325

Area = s^2, where s = length of the side of the square, 327

Rectangle

Perimeter = **2l + 2w**, where l = length of the rectangle and w = width of the rectangle, 325

Area = **lw**, where l = length of the rectangle and w = width of the rectangle, 327

Parallelogram

Perimeter = **2b + 2s**, where b = the base and s = length of the adjacent side, 325

Area = **bh**, where b = length of the base and h = height of the parallelogram, 327

Triangle

Perimeter = **a + b + c**, where a, b, and c are the lengths of the sides of the triangle, 325

Area = $\frac{1}{2}$**bh**, where b = base of triangle and h = height of triangle, 327

*Page references are included when appropriate.

Area of a triangle if the *height* is not known (Heron's formula):

Area $= \sqrt{s(s-a)(s-b)(s-c)}$, where $s = \frac{1}{2}(a+b+c)$ and $a =$ side 1, $b =$ side 2, and $c =$ side 3 of the triangle, 328

Trapezoid

Perimeter $= a + b + c + d$, where a, b, c, and d are the lengths of the sides, 325

Area $= \frac{1}{2}h(b_1 + b_2)$, where $h =$ height of trapezoid, $b_1 =$ length of shortest leg, and $b_2 =$ length of longest leg, 327

Circle

Circumference $= 2\pi r$ or πd, where $r =$ radius and $d =$ diameter, 330

Area $= \pi r^2$, where $r =$ radius, 330

Area of a sector $= \left(\frac{\theta}{360°}\right) \cdot \pi r^2$, where $\theta =$ angle (measured in degrees) that forms the sector and $r =$ radius of circle

Area of a sector $= \frac{\theta r^2}{2}$, where $\theta =$ angle (measured in radians) that forms the sector and $r =$ radius of circle

Length of an arc $= \frac{\pi r \theta}{180}$, where $\theta =$ angle (measured in degrees) that forms the arc and $r =$ radius of circle

Length of an arc $= r\theta$, where $\theta =$ angle (measured in radians) that forms the sector and $r =$ radius of circle

Area and Volume of Three-Dimensional Objects
Cube

Lateral Surface Area $= 4s^2$, where $s =$ length of one side of cube

Total Surface Area $= 6s^2$, where $s =$ length of one side of cube

Volume $= s^3$, where $s =$ length of one side of cube

Right Rectangular Prism

Lateral Surface Area = $(2l + 2w)h$, where l = length of prism, w = width of prism, and h = height of prism, 338

Total Surface Area = $(2l + 2w)h + 2lw$, where l = length of prism, w = width of prism, and h = height of prism, 338

Volume = lwh, where l = length of prism, w = width of prism, and h = height of prism, 338

Right Circular Cylinder

Lateral Surface Area = $(\pi d)h$, where d = diameter of circular base and h = height of cylinder, 338

Total Surface Area = $(\pi d)h + 2\pi r^2$, where d = diameter of circular base, r = radius of circular base, and h = height of cylinder, 338

Volume = $\pi r^2 h$, where r = radius of circular base and h = height of cylinder, 338

Pyramid

Volume = $\frac{1}{3}Bh$, where B = area of the polygon base and h = height of the pyramid

Right Circular Cone

Lateral Surface Area = πrs, where r = radius of the circular base and s = slant height of cone, 338

Total Surface Area = $\pi rs + \pi r^2$, where r = radius of the circular base and s = slant height of cone, 338

Volume = $\frac{\pi r^2 h}{3}$, where r = radius of the circular base and h = height of the cone, 338

Sphere

Total Surface Area = $4\pi r^2$, where r = radius of the sphere, 338

Volume = $\frac{4}{3}\pi r^3$, where r = radius of the sphere, 338

Special Triangle Formulas and Relationships

Pythagorean theorem (for right triangles where c is the hypotenuse and a and b are the legs) $c^2 = a^2 + b^2$, 353

30°, 60°, 90° right triangle relationship among the sides

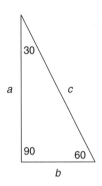

$a = b\sqrt{3}$, where a = the side opposite 60° angle and b = the side opposite 30° angle, 357

$c = 2b$, where c = the hypotenuse and b = the side opposite 30° angle, 357

45°, 45°, 90° right triangle relationship among the sides

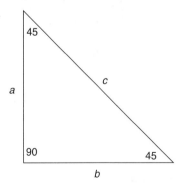

$a = b$, where a and b are the legs of the triangle, 355

$c = a\sqrt{2}$, where c = the hypotenuse and a = one of the legs of the triangle, 355

■ A-5 Business Formulas

Daily Simple Interest for $100

$I = P\left(\dfrac{R \times D}{365}\right)$, where I = simple exact interest, P = the principal, R = the given annual interest rate (expressed as a decimal), D = the given number of days

Future Value of a Compounded Lump Sum Investment

$FV = P(1 + R)^N$, where FV = future value, P = principal, R = interest rate per period (expressed as a decimal), N = total number of compounding periods, 269

Compound Interest of a Lump Sum Investment

$I = P(1 + R)^N - P$, where I = compound interest, P = principal, R = interest rate per period (expressed as a decimal), N = total number of compounding periods, 268

Present Value of a Compounded Lump Sum Investment

$PV = \dfrac{FV}{(1 + R)^N}$, where PV = present value, FV = future value, R = interest rate per period (expressed as a decimal), N = total number of periods, 270

Effective Interest Rate

$E = \left(1 + \dfrac{r}{n}\right)^n - 1$, where E = effective rate, r = interest rate per year, and n = number of compounding periods per year, 271

Future Value of a Lump Sum Investment Compounded Continuously

$FV = Pe^{rt}$, where FV = future value, P = principal, r = rate per year, and t = time in years, 273

Future Value of an Ordinary Annuity

$FV = P\left[\dfrac{(I + R)^N - 1}{R}\right]$, where FV = future value, P = amount of periodic payment, R = interest rate per period (expressed as a decimal), N = total number of periods, 274

Present Value of an Ordinary Annuity

$A = \dfrac{(1 + R)^N - 1}{R \times (1 + R)^N}$, where A = present value of the ordinary annuity, R = rate per period (expressed in decimal form), and N = total number of periods

Sinking Fund Payment

$P = FV\left[\dfrac{R}{(1 + R)^N - 1}\right]$, where P = sinking fund payment, FV = future value, R = interest rate per period (expressed as a decimal), and N = total number of periods, 277

Monthly Payment for an Amoritized Loan

$M = P\left[\dfrac{R}{1 - (1 + R)^{-N}}\right]$, where M = monthly payment, P = principal or initial amount of the loan, R = interest rate per month, and N = total number of months, 278

■ A–6 Conversion Factors

U.S. CUSTOMARY	To Change		
	From	To	Multiply By
Length or *Distance*			
12 inches (in.) = 1 foot (ft)	feet	inches	12
	inches	feet	0.833333
3 feet (ft) = 1 yard (yd)	feet	yards	3
	yards	feet	0.333333
36 inches (in.) = 1 yard (yd)	inches	yards	36
	yards	inches	0.027778
5280 feet (ft) = 1 mile (mi)	feet	miles	5280
	miles	feet	0.000189

Weight or Mass

16 ounces (oz) = 1 pound (lb)	ounces	pounds	16
	pounds	ounces	0.0625
2000 pounds (lb) = 1 ton (T)	pounds	tons	2000
	tons	pounds	0.0005

Liquid Capacity or Volume

3 teaspoons (t) = 1 tablespoon (T)	teaspoon	tablespoon	0.333333
	tablespoon	teaspoon	3
2 tablespoons (T) = 1 ounce (oz)	tablespoon	ounce	0.5
	ounce	tablespoon	2
8 ounces (oz) = 1 cup (c)	ounces	cups	8
	cups	ounces	0.125
2 cups (c) = 1 pint (pt)	cups	pints	2
	pints	cups	0.5
2 pints (pt) = 1 quart (qt)	pints	quarts	2
	quarts	pints	0.5
4 quarts (qt) = 1 gallon (gal)	quarts	gallons	4
	gallons	quarts	0.25

METRIC SYSTEM	From	To	Multiply By

Length or Distance

1 kilometer (km) = 1,000 meters (m)	kilometers	meters	1,000
	meters	kilometers	0.001
1 hectometer (hm) = 100 meters	hectometers	meters	100
	meters	hectometers	0.01
1 dekameter (dkm) = 10 meters	dekameters	meters	10
	meters	dekameters	0.1
1 decimeter (dm) = 0.1 meter	decimeters	meters	0.1
	meters	decimeters	10
1 centimeter (cm) = 0.01 meter	centimeters	meters	0.01
	meters	centimeters	100
1 millimeter (mm) = 0.001 meter	millimeters	meters	0.001
	meters	millimeters	1,000

Weight

1 kilogram (kg) = 1000 grams (g)	kilograms	grams	1,000
	grams	kilograms	0.001
1 hectogram (hg) = 100 grams	hectograms	grams	100
	grams	hectograms	0.01
1 dekagram (dkg) = 10 grams	dekagrams	grams	10
	grams	dekagrams	0.1
1 decigram (dg) = 0.1 gram	decigrams	grams	0.1
	grams	decigrams	10

1 centigram (cg) = 0.01 gram	centigrams	grams	0.01
	grams	centigrams	100
1 milligram (mg) = 0.001 gram	milligrams	grams	0.001
	grams	milligrams	1,000

Capacity

1 kiloliter (kL) = 1,000 liters (L)	kiloliters	liters	1,000
	liters	kiloliters	0.001
1 hectoliter (hL) = 100 liters	hectoliters	liters	100
	liters	hectoliters	0.01
1 dekaliter (dkL) = 10 liters	dekaliters	liters	10
	liters	dekaliters	0.1
1 deciliter (dL) = 0.1 liter	deciliters	liters	0.1
	liters	deciliters	10
1 centiliter (cL) = 0.01 liter	centiliters	liters	0.01
	liters	centiliters	100
1 milliliter (mL) = 0.001 liter	milliliters	liters	0.001
	liters	milliliters	1,000

U.S. CUSTOMARY AND METRIC COMPARISONS	From	To	Multiply By
Length			
1 meter = 39.37 inches	meters	inches	39.37
	inches	meters	0.0254
1 meter = 3.28 feet	meters	feet	3.2808
	feet	meters	0.3048
1 meter = 1.09 yards	meters	yards	1.0936
	yards	meters	0.9144
1 centimeter = 0.39 inch	centimeters	inches	0.3936
	inches	centimeters	2.54
1 millimeter = 0.04 inch	millimeters	inches	0.03937
	inches	millimeters	25.4
1 kilometer = 0.62 mile	kilometers	miles	0.6214
	miles	kilometers	1.6093
Weight			
1 gram = 0.04 ounce	grams	ounces	0.0353
	ounces	grams	28.3495
1 kilogram = 2.2 pounds	kilograms	pounds	25.2046
	pounds	kilograms	0.4536
Liquid Capacity			
1 liter = 1.06 quarts	liters	quarts	1.0567
	quarts	liters	0.9463

■ A–7 Metric Prefixes*

Prefix	Symbol	Times the Standard Unit
Yotta-	Y	10^{24}
Zetta-	Z	10^{21}
Exa-	E	10^{18}
Peta-	P	10^{15}
Tera-	T	10^{12}
Giga-	G	10^{9}
Mega-	M	10^{6}
kilo-	k	10^{3}
hecto-	h	10^{2}
deka-	dk	10^{1}
Standard Unit		10^{0}
deci-	d	10^{-1}
centi-	c	10^{-2}
milli-	m	10^{-3}
micro-	μ	10^{-6}
nano-	n	10^{-9}
pico-	p	10^{-12}
femto-	f	10^{-15}
atto-	a	10^{-18}
zepto-	z	10^{-21}
yocto-	y	10^{-24}

*The metric conversion tables can be extended to include the *very large* and *very small* metric measures by using the prefixes given in the Metric Prefixes table.

■ A–8 More Conversion Factors

	From	To	Multiply By
UNITS OF TIME			
1 minute = 60 seconds	minutes	seconds	60
	seconds	minutes	0.01667
1 hour = 60 minutes	hours	minutes	60
	minutes	hours	0.01667
1 day = 24 hours	days	hours	24
	hours	days	0.04167
1 week = 7 days	weeks	days	7
	days	weeks	0.1429
1 fortnight = 2 weeks	fortnights	weeks	2
	weeks	fortnights	0.5

1 month = 30 days (ordinary time)	months	days	30
	days	months	0.03333
1 leap month = 29 days	leap months	days	29
	days	leap months	0.0345
1 year = 12 months	years	months	12
	months	years	0.08333
1 year = 365 days	years	days	365
	days	years	0.0027
1 decade = 10 years	decades	years	10
	years	decades	0.1
1 century = 100 years	centuries	years	100
	years	centuries	0.01

UNITS OF AREA

1 square foot = 144 square inches	square feet	square inches	144
	square inches	square feet	0.0069
1 square yard = 9 square feet	square yards	square feet	9
	square feet	square yards	0.1111
1 square mile = 2.5887 square kilometers	square miles	square kilometers	2.5887
	square kilometers	square miles	0.3863

UNITS OF VOLUME (CAPACITY)

1 cubic inch = 16.387 cubic centimeters	cubic inches	cubic centimeters	16.387
	cubic centimeters	cubic inches	0.0610
1 cubic inch = 0.01639 liters	cubic inches	liters	0.01639
	liters	cubic inches	61.0128
1 cubic foot = 0.0283 cubic meter	cubic feet	cubic meters	0.0283
	cubic meters	cubic feet	35.3357
1 teaspoon = 4.93 milliliters	teaspoons	milliliters	4.93
	milliliters	teaspoons	0.2028
1 tablespoon = 14.97 milliliters	tablespoons	milliliters	14.97
	milliliters	tablespoons	0.0668

1 fluid ounce = 29.57 milliliters	fluid ounces	milliliters	29.57
	milliliters	fluid ounces	0.0338
1 cup = 0.24 liters	cups	liters	0.24
	liters	cups	4.1667
1 pint = 0.47 liters	pints	liters	0.47
	liters	pints	2.1277
1 gallon = 0.00379 cubic meters	gallons	cubic meters	0.00379
	cubic meters	gallons	263.85

■ A–9 Changing Temperature between Fahrenheit and Celsius

Fahrenheit to Celsius Subtract 32 then multiply by $\frac{5}{9}$

Celsius to Fahrenheit Multiply by $\frac{9}{5}$ then add 32

Or use the formulas:

$$C = \frac{5}{9}(°F - 32) \qquad F = \frac{9}{5}C + 32$$

■ A–10 Special Algebra Patterns for Factoring

$a^2 + 2ab + b^2 = (a + b)^2$

$a^2 - b^2 = (a + b)(a - b)$

$a^3 + b^3 = (a + b)(a^2 - ab + b^2)$

$a^3 - b^3 = (a - b)(a^2 + ab + b^2)$

■ A–11 Using Technology

There are many mathematical tools that use technology to assist in mathematical calculation and problem solving. There are many models and brands of calculators, handheld computers, and computer software. Each one will have different features and different procedures for accessing these features. The ideal situation is to have the user's manual handy whenever you need to use a new feature. However, this is not always the case. It is important to develop investigative strategies for determining the proper sequence for performing calculations, drawing graphs, making tables, and using other technological features.

The keystroking conventions that are required to perform a calculation are often referred to as **syntax.** When a calculator or program indicates that there is a **syntax error,** it means that a mistake has been made in the way the steps have been entered. Some calculators or programs will indicate where the first syntax error occurred by giving a step number or by moving the cursor to that location. This will help you to identify and to correct errors.

Some of the more advanced calculator features may require that you get help from an experienced user or that you refer to your user's manual. However, most of the more commonly used features can be learned by using investigative strategies.

Types of Calculator and Computer Tools

The most common types of calculators are basic, scientific, financial, graphing, and office. The **basic calculator** is the least expensive and most common. It performs the basic operations of addition, subtraction, multiplication, and division and often has a square root key and memory keys. To perform a series of calculations on a basic calculator, the result of a step is recorded and reentered when needed or it is temporarily stored in memory. **Office calculators** often have an electronic display of the number currently being entered or the result of a calculation as well as a list of the numbers and calculations printed on paper. Like the basic calculator, they have the fewest features but are the easiest to learn to use.

The **scientific, business,** and **financial calculators** have many additional functions, numerous storage locations, and can make a series of calculations by using the **order of operations.** Many of these calculators will have two lines on the display. One line shows the series of keystrokes that have been entered so that you can proof and correct any errors. The other line shows the result of the calculations. The **graphing calculator** has a multi-line display screen with many editing capabilities and features for drawing graphs and making data tables.

The **electronic spreadsheet,** like Excel®, Lotus®, and Quattro Pro®, is a computer program that displays information in rows and columns of a table. Calculations are made by writing the steps of the calculations in formulas using the order of operations and mathematical concepts.

Order of Operations

With either a calculator or spreadsheet, the standard order of operations is presumed (parentheses or groupings, exponents [powers and roots], multiplication and division, and addition and subtraction). The only available grouping symbol is parentheses so all other types of groupings (braces, brackets, bar, etc.) are translated to parentheses. Other conventions vary from calculator to calculator.

Most calculators except the basic and office calculators have **parentheses keys** for facilitating the entry of a series of calculations. Some require a times sign before the parentheses, such as $3 \times (5 + 7)$, and others do not, $3(5 + 7)$. The result of these calculations is 36. To determine how your calculator works, *always start with a problem you can work mentally and try entering the problem several ways.* Often you will find that there is more than one way to get a correct solution. Even *try series that you don't expect to be correct to become familiar with the calculator's limitations.* Try $3 \times 5 + 7$. The result of these calculations is 22. An option that does not require parentheses is to enter the problem in steps. For $3 \times (5 + 7)$, complete the addition first then multiply the sum. Enter $5 + 7 =$ and 12 will be displayed as the sum. Then, enter $12 \times 3 =$ and 36 will be displayed as the final result of the calculations.

To perform a long series of calculations as in many of the busines formulas, great care must be used to enter the series correctly. The series of calculations necessary to find the future value of an annuity due (p. 275) requires careful use of parentheses inside of parentheses.

EXAMPLE (from p. 275)   To find the future value of a quarterly annuity due of $100 for five years at 5% compounded quarterly:

$$FV = 100\left[\frac{(1 + 0.0125)^{20} - 1}{0.025}\right][1 + 0.0125]$$

$100\boxed{\times}\boxed{(}\boxed{(}\boxed{(}\boxed{(}1 +\boxed{.}\boxed{0125}\boxed{)}\boxed{\wedge}\boxed{20} - 1\boxed{)}\boxed{\div}\boxed{.}\boxed{0125}\boxed{)}\boxed{\times}\boxed{(}1 +\boxed{.}\boxed{0125}\boxed{)}\boxed{\text{Enter}} \Rightarrow 2284.501577$

We can set up an electronic spreadsheet to evaluate formulas for different values of the variable. Each value of each variable can be assigned a **location** or **address** called a **cell**. The formula is written using the variable address.

In the formula for the future value of an annuity:

$$FV = P\left[\frac{(1 + R)^N - 1}{R}\right][1 + R],$$

assign P to cell C3, R to cell C4, and N to cell C5. Then write the formula for C7 as:

= C3*(((1 + C4)^C5 − 1)/C4)*(1 + C4), where $\wedge$ is used to indicate powers and / is used to indicate division.

Then, as different values are entered in the variable cells, the formula results cell C7 will show the results of the calculations.

∟

Built-In Functions

Many calculators and computer programs have common business and statistical formulas already programmed. Then, the user will only have to access the function through menu options or keystroke options and enter the variable values. These built-in functions require that the variable values be entered a certain way or using a prescribed **syntax.** The following quote is from the Microsoft Excel® help file:

FV(rate,npr,pmt,pv,type)

For a more complete description of the arguments in FV and for more information on annuity functions, see PV.

Rate   is the interest rate per period.

Nper   is the total number of payment periods in an annuity.

Pmt   is the payment made each period; it cannot change over the life of the annuity. Typically, pmt contains principal and interest but no other fees or taxes. If pmt is omitted, you must include the pv argument.

Pv   is the present value, or the lump-sum amount that a series of future payments is worth right now. If pv is omitted, it is assumed to be 0 (zero), and you must include the pmt argument.

Type   is the number 0 or 1 and indicates when payments are due. If type is omitted, it is assumed to be 0.

Set type equal to	If payments are due
0	At the end of the period (ordinary annuity)
1	At the beginning of the period (annuity due)

Remarks

- Make sure that you are consistent about the units you use for specifying rate and nper. If you make monthly payments on a four-year loan at 12 percent annual interest, use 12%/12 for rate and 4*12 for nper. If you make annual payments on the same loan, use 12% for rate and 4 for nper.

- For all the arguments, cash you pay out, such as deposits to savings, is represented by negative numbers; cash you receive, such as dividend checks, is represented by positive numbers.

The syntax for finding the future value using the built-in function in Excel® is:

$$= FV(0.0125, 20, -100, 0, 1)$$

Similar functions are built into financial or graphing calculators. Each calculator model and brand has its own syntax. For example, the TI 83 Plus® has the following syntax:

Access the menu from the blue APPS key.

Then select Option 1: Finance by pressing ENTER.

Access option 6 by pressing 6.

Enter the values of the problem using the following syntax.

$(N, I\%, PV, PMT, P/Y, C/Y)$ where

N = total number of payments

$I\%$ = annual interest rate

PV = present value (zero when finding future value)

PMT = payment amount (enter cash outflows as negative numbers)

P/Y = number of payment periods per year

C/Y = number of compounding periods per year

The syntax for finding the future value using the built-in function in TI–83 Plus® is:

tmv_FV(20,5,0,−100,4,4)

Glossary/Index